Integrated Urban Environment Management and Resilience

SCIENCES

Architecture, Urban Planning and Development,
Field Director – Patrizia Laudati

Integrated Environment Management and Resilience,
Subject Heads – Marc Méquignon and Jean-Pierre Mignot

Integrated Urban Environment Management and Resilience

From the Contemporary City to Sustainable Urbanity

Coordinated by
Luc Adolphe

iSTE

WILEY

First published 2022 in Great Britain and the United States by ISTE Ltd and John Wiley & Sons, Inc.

ISTE Ltd
27-37 St George's Road
London SW19 4EU
UK

www.iste.co.uk

John Wiley & Sons, Inc.
111 River Street
Hoboken, NJ 07030
USA

www.wiley.com

Library of Congress Control Number: 2022940281

British Library Cataloguing-in-Publication Data
A CIP record for this book is available from the British Library
ISBN 978-1-78945-077-4

ERC codes:
SH2 Institutions, Values, Environment and Space
SH2_6 Sustainability sciences, environment and resources
SH2_8 Energy, transportation and mobility
SH2_9 Urban, regional and rural studies
SH2_10 Land use and regional planning
SH2_11 Human, economic and social geography

Contents

Introduction

Jean-Pierre MIGNOT and Marc MÉQUIGNON
LERASS, UPS, Toulouse, France

The city, in particular urbanization and urbanity in general, kindles understandable interest vis-à-vis the preferred models that are adopted (chosen or mandated) so as to better occupy our common spaces, the planet, without destroying it. If current projection trends hold true over the next century, nearly 80% of the planet's population will end up living in cities, that is, urban systems. Of course, and we will never stop repeating it, the city is an artifact, a technical whole, a constructed technical system, a sentiment Ellul (1977) would have no doubt echoed; however, it has also been analyzed and perceived as a societal problem, further emphasizing the importance and pivotal role it has in shaping future developments. It is for this reason that we have deliberately chosen a multidisciplinary analysis of the city and the problems facing it.

Under these conditions, the problematic city that we propose is best viewed through a double lens: at once as an artifact, that is, a more or less homogeneous technical whole that reflects the system within which it is deployed, and also as a production of space and a space for production, that is, a privileged site that hosts the unfolding of social relations throughout all of its historical forms, which overlap, and within which it develops. Since it is permeated by all forms of sociality, and carries all the contradictions and ills of society (but also all of its joys) within it, the city is directly subjected to and even constitutes this privileged vector. As such, it is easy to comprehend the reasons why we approach problematic urbanity through a deliberately multidisciplinary perspective, which naturally leads us to a fresh perspective, multiple and rich in tangible determinations: we wholeheartedly believe that the illumination and understanding of the city, as proposed here, can only be extensive.

Furthermore, in light of the modern rhetoric on global warming and the ecological transition that has become all but consubstantial, the modern concept of the city is shifting towards a "sustainable" city, which is not only particularly important, but also makes obvious reference to the numerous conflicts with the environment that the city has incited, and moreover reveals the desire to create the conditions for a more serene future than the one promised to us, should we continue to occupy the planet (in general), and the city in particular, in the same way as has been done for too many decades now. This relatively recent concept of the sustainable city appears to stem from a double requirement, which we have observed, at least on the surface:

– on the one hand, to introduce into "the lived social world" the technical developments systematically associated with economic and societal progress, which are likely to modify (and, for some, improve) the conditions of daily life, both in country towns and cities alike, for generations to come;

– on the other hand, how it has become a determining factor of sustainability that will ensure the harmonious development of the planet, and by consequence the cities that compose it, which are themselves (but not exclusively) the cause behind the general disturbances that we are suffering from, because the required development necessitates destructive conditions, and as a result, constrains us.

It is through this perspective of conflict, between "technical progress" and the endangerment of the planet by human activity, on the one hand, and on the other hand, the systematic deployment of these same techniques that the concept of the smart city seems to impose, wherein technology at the service of humanity thus becomes the "*deus ex machina*" of a new world. The city appears to be a highly technical artifact that serves both as a basic substrate, capable of fostering "the bright future of humanity", and at the same time constitutes the main vector for the sustainability of the planet, with the ability to shelter people for generations to come. Moreover, this expression (before becoming a concept) finds equivalent expressions (or concepts) in the recent past, such as "the ring road", and it is also this semantic change that we wish to account for. Therefore, the new city at the service of a humanity that tries to protect itself against that which it has itself contributed to developing, becomes a determining axis through which it is possible to think about the city of tomorrow, as well as the development of management tools in service of its sustainability. We thus propose to think of the city as a societal issue, on the one hand, and on the other hand, under the form of an encyclopedic approach, that is as a whole comprised of many definitions, under the guidance of authors whose work synthesizes the many challenges of implementation. The different aspects addressed in this framework can be summarized as follows:

– an approach to the city through the presentation of morpho-sociological analysis, in particular with reference to works by Benevolo, Choay, Riboulet, etc., whose objective is to represent the state of the contemporary city and the conditions of its production;

– an economic approach concerning, in particular, the exact modalities of a specific development model, which sees urban composition as the key to putting the paradoxes of the smart city into perspective as a possible response to the problems pertaining to the sustainability of the city;

– a geographical approach to the smart city, in particular by addressing the problems facing development zones, and the worldwide nature that shares a need for this type of solution;

– a sociological approach to the city as a place of production of modernity, in all its forms, as an instance of rationalization that leads from individual to social differentiation.

In order for the proposed analysis to take into account a plural vision of the city, we call upon a multidisciplinary approach. Accordingly, Chapter 1, by Marc Méquignon and Patrizia Laudati, sets itself the objective of providing critical elements for an examination of the city through a morpho-sociological approach. The city is viewed here as a complex and moving object of study, requiring by its very definition, an approach that goes beyond disciplinary partitions and schools of thought, outside time and space, so as to try and grasp the existing connections between all the urban components, objects and subjects, built and non-built elements, urban practices, meanings, etc.; the whole as a result of its modes of production. A purely syntactic or morphological approach to the urban would, on its own, would not suffice to grasp what Berque terms “urbanity”, namely: “this more general composition where spatial entities and social entities come into resonance” (Berque (1993), *author’s translation*). Obviously, we are not defending a “standard” model of urbanity, but rather studying the different elements that make up urbanity, which, far from being invariable, base their differences on specific spatial, social and cultural structures.

A morpho-sociological approach, focusing on both spatial entities (their form(s)) and social entities (uses, practices, meanings), as such seems more pertinent for demonstrating the inextricable interweaving between these two dimensions, that is, between the materiality and spirit of a city. In addition, any analysis in the service of a more in-depth knowledge of urbanity will make it possible to respond in a more relevant way to the needs and requirements of urban life, by means of the elements identified. These operational elements can then be incorporated into development policies that will help to rethink the urban project.

First, by borrowing Riboulet's work on the four modes of production of the city (traditional, princely, liberal and regulatory), we emphasize how, in the constructed and lived city, these four modes inevitably coexist (Riboulet 1998). This testifies to the complexity and difficulty of apprehending our multidimensional object of study (the city), which we question both as an object and as a subject.

Second, we propose possible reading keys, which, in a systemic and circular logic, lead to the development of the various elements composing urbanity: we thus introduce the concepts of urban forms (both forms in the plural and in the singular), in order to understand them through their constituent elements: structure, built and non-built elements, singular elements, disruptions, materials, techniques and technologies, zones, real estate markets and social flows, renovation, development, and representations (drawing and design). Each time, no matter the element under study, it is always understood as a triptych, that is, by including its morphology, uses and meanings.

A city is never the result of chance. It arises from a need and adapts as the society that hosts and shelters within it evolves:

> The city – place of organized settlement, seat of authority, arises from the village, but not just from the swollen village. It is formed when industries and services are no longer exercised by the people who cultivate the land, but by others who do not have this obligation and are maintained by the former thanks to a surplus of the total production (Benevolo (1995, p. 13), *author's translation*).

Therefore, we can say that the city arises from an opposition between two social groups, the first of which is dominant over the other, which is subordinate. The spatial forms are thus a reflection of these power games. The mode of urban composition corresponds to the social processes that have been put into action, with the aim of producing an inhabited space, comprising all the functions seen as useful to it at that given moment in history, imbuing within the same movement, a form and meaning to the space. It is thus also a representation of the power and social relations that promote it.

In Chapter 2, the developments proposed by Jean-Pierre Mignot, through the lens of an economic approach, deliberately make the decision to not give in to an "orthodox" interpretation typical to neoliberal economic analysis, but rather seek to answer the broader question of the city's economic space by viewing it as a space for the production, distribution and consumption of wealth it has produced, and also, and just as importantly, as a production of this space within which these economic activities take place. Within this context, the question of space and the meaning given to it helps us to better understand the scope of the duality that characterizes the city: as a space for production and as a production of this space. The technical forms of the city, as well as its social (and/or societal) forms, are deployed in time (and therefore in history) in tandem with the space which hosts it. The author thus communicates the perspective that the emergence of a specific space that belongs to history assigns a history for the production of this space, as well as the history of this same space at the same time. By favoring the forms of wealth accumulation for time and space, we can therefore distinguish the forms that the city will take, that is, the different forms that the conditions of production will give to the city as an expression of the social production and exchange within which it is developing. For example, the Fordist city is therefore an instance in which, on the one hand, the car is introduced into the cityscape up until the point of polluting it, and on the other hand, the production conditions of Fordist accumulation led society, as of the 1970s, to denounce the infamous "Meadows Report" (better known as the "Club of Rome" report). The birth of the sustainable city is the result of an awareness that changes of this nature, should they persist, may lead to a worst-case scenario. Similarly, the "neoliberal city" incorporates this logic of accumulation, but fails to respond to the problems responsible for environmental damage, brought about by the emergence of agglomerations whose exploding populations often exceed 20 million inhabitants. As such, the sustainable city becomes the beacon of hope on the horizon, in contrast to a vantage that predicts a planet that will engulf itself, should it continue along a path fatal to its own future.

To approach the problems of the modern city without an expression that only geography can provide is, in our minds, heresy. To this end, Chapter 3, by Fabrice Escaffre, Mélanie Gambino and Sinda Haouès-Jouve, offers an approach to urban spaces based on issues of sustainability and habitability. After analyzing the general dynamics that characterize contemporary urbanization while recalling the diversity of urban spaces, as seen through four facets, considered here as four interfaces, these facets are discussed in greater detail. The first facet deals with housing and habitat, tackling the classical topic most often considered in the analysis of cities. It is an opportunity to highlight the gains of an approach to cities by examining the transversal nature of inhabitant practices. The second facet examines mobility, framing urban spaces through the multiple flows that animate them and give them shape. These first two facets allow us to focus on the social inequalities that characterize cities, as much as on the environmental issues that accompany them. The third facet shifts the analysis towards a subject a priori considered as only remotely connected to urban issues, but which is being integrated more and more strongly, namely agriculture. Whether this is explained as an extension of cities to their rural environment or as a result of food supply concerns, urban agriculture is being treated as a new angle to the urban approach. Finally, climate serves as a focal point for the final facet, as it implies a shift in focus from the local to the global, especially given that the multiple urban effects brought about by its change influence the three previous facets. Through the intermediation of climate, the question of the sustainable city and its adaptability to new living conditions that climate change itself imposes on our way of living and understanding of the city, the city is thus presented in a transformed way.

Finally, stemming from an essentially sociological literature, Chapter 4 describes the city as a space for the production of modernity, that is, for the production of rationalization, individualization and differentiation of societies. Rationalization is embodied in the city as human enterprise, harnessing the environment and organizing its way of life. Individualization describes a city dweller who has freed themself from the grip of the collective yoke peculiar to traditional societies, and claims to be the ruler of their own existence. Differentiation accounts for the growing complexity of society, which is fragmenting into self-referenced subsystems. The fruit and vector for the dynamics of modernity, the city is also the theater of its tensions. The city generates a so-called "urban" society, which results in a redefinition of the modes of sociability based on a difficult-to-balance equation: the making of a society, which produces solidarity alongside emancipated, de-rooted, multi-belonging and sometimes disengaged individuals. The seat of personal freedom, the city indeed produces a "specific mentality" that reshapes the relationships and social aspirations of individuals. The city is in this sense the space of secret, reserved, rational relations, of elective sociability, generating a progressive dislocation of community-type social bonds. This city of individuals questions the propensity of city dwellers to accept the constraints that life within society implies, give themselves a collective plan, and decenter and tolerate differences. Urban planners and architects thus attempt to determine sociability in their projects, while initiatives from civil society are undertaken to invent new forms of committing to the common good. However, these initiatives must also confront deep social logics in favor of the social homogenization that is imprinted upon urban space. Urban society is made up and broken down according to socio-spatial population dynamics, which reflect the stratification of society into classes that are guided by mechanisms of distinction. Public policies and proactive approaches aimed at building social ties between different classes are multiplying without overwhelmingly succeeding in reversing the tendencies towards the social homogenization of urban spaces. The city is fragmented from above and below, following sociological and economic constraints, urban space is torn apart, segregated and unraveled. The result is a concentration of social inequalities which has a moral impact on all city dwellers, whatever their social position. Social diversity is a horizon that seems to have been reached; however, an intellectual, even utopian, horizon, runs against the spontaneous, unspeakable or unconscious desire of those who resemble one another to unite. The sustainable city, the latest mutation of the urban imagination, simultaneously concentrates the aspirations and contradictions of urban society. It is intended as a response to the ecological and social crises affecting modernity, and also as a part of a globalized economic competition in which cities are prominent players. In the court of injunctions to sustainability, the city is accused of all evils: pollution, poverty, inequality; and is also perceived as the space for remediation. The initiation of profound transformations within urban society could inaugurate a new, more responsible, more equitable world. Once again, this desire for renewal comes up against conflicting social representations and aspirations of city dwellers, who want both the city and the countryside. The sustainable city seems to imply certain cessations; however, the individual home continues to embody the outcome of a successful residential trajectory. The sustainable city is also conceived and built within a new paradigm, which replaces the techno-scientific references, and at the same time a regime of uncertainty, a so-called "theoretical cooling" is developed. The urban translation that questions the progressive ideology brought about by the industrial revolution is embodied in the concept of the urban project. It promotes an empirical and partnership approach intended to experiment with local choices that are developed collectively. But then again, the solutions it produces also generate exclusion, segregation and labeling.

Obviously, the rules imposed on us by adopting a multidisciplinary approach implies that, ideally, we should have invited other voices, whether they belong to the humanities and social sciences (HSS) or the "hard" sciences. However, it was a deliberate choice to only promote the expression of HSS, for the sake of homogenizing discourse and a desire to understand the city from this unique angle. Not that the opinions of our colleagues in the "hard" sciences are not welcome, but the objective of the editors and the available space gave us little room to maneuver. We would like to thank each and every one for their contribution, with the hopes that we have provided our readers with the opportunity to better understand the issues concerning the future of cities and, through them, the future of the planet.

References

Benevolo, L. (1995). *Histoire de la ville*. Parenthèses, Marseille.

Berque, A. (1993). *Du geste à la cité. Formes urbaines et lien social au Japon*. Gallimard, Paris.

Ellul, J. (1977). *Le Système technicien*. Calmann-Lév, Paris.

Riboulet, P. (1998). *Onze leçons sur la composition urbaine*. Presses de l'École nationale des ponts et chaussées, Paris.

1

Morpho-Sociological Approach to the City

Marc Méquignon[1] and Patrizia Laudati[2]
[1]*LERASS, UPS, Toulouse, France*
[2] *SicLab Méditerranée, UCA Université Côte d'Azur, Nice, France*

1.1. Introduction: socio-morphological analysis as a tool for understanding the city

The objective of this chapter is to provide critical elements for the analysis of the city through a morpho-sociological approach. The city is a complex and moving object of study requiring, by its very definition, an approach that goes beyond disciplinary partitions and schools of thought, outside of time and space, in order to try and grasp the existing connections between all the urban components, objects and subjects, built and non-built elements, urban practices, meanings, etc., the whole a result from its modes of production.

A purely syntactic or morphological approach to that which is considered urban would not suffice on its own, if we are to grasp what Berque calls urbanity, namely: "this more general composition where spatial entities and social entities enter into resonance" (Berque (1993), author's translation).

Of course, we are not defending a "standard" model of urbanity, but rather an interpretation of the elements that make up urbanity which, far from being invariable, base their differences on specific spatial, social and cultural structures.

Integrated Urban Environment Management and Resilience,
coordinated by Luc Adolphe.

A morpho-sociological approach, focusing on both spatial entities (their form/s) and social entities (uses, practices, meanings), therefore seems more relevant to show the inextricable interconnectedness between these two dimensions, between the materiality and spirit of a city.

Moreover, any analysis that serves a more in-depth knowledge of urbanity will make it possible to respond in a more relevant way to the requirements and needs of urban life, through the elements that have been identified. These operational elements can then be incorporated into development policies to rethink the urban project.

First, by borrowing Riboulet's work on the four modes of production of the city (traditional, princely, liberal and regulatory), we emphasize how, in the constructed and lived city, these four modes inevitably coexist (Riboulet 1998). This testifies to the complexity and difficulty of apprehending our multidimensional object of study (the city), which we question both as an object and as a subject.

In a second step, we propose possible reading keys which, in a systemic and circular logic, lead us to develop the different elements composing urbanity: we introduce the concepts of urban forms (both plural and singular forms), in order to understand them through their constitutive elements: structure, built and non-built elements, singular elements, ruptures, materials, techniques and technologies, parcels of land, real estate markets and social flows, renovation, planning, representations (drawing and design). Each time, no matter the element being studied, it is always understood as a triptych that includes its morphology, uses and meanings at the same time.

A city is never the fruit of chance. It is born of needs and adapts as it evolves:

> The city – a place of organized settlement, the seat of authority – is born of the village, but is not just the enlargement of a village. It is formed when industries and services are no longer exercised by the people who cultivate the land, but by others who do not have this obligation and are maintained by the former by the surplus of the total production (Benevolo (1995, p. 13), author's translation).

Therefore, we can say that the city is thus created from the opposition of two social groups, the first is dominant and the other is subordinate. The spatial forms are thus the reflection of these power games. The mode of urban composition corresponds to the social processes put into action with the aim of producing an inhabited space that includes all the functions seen as useful at that given moment in history, imbuing within the same movement a form and meaning to the space.

1.2. The city's modes of production

This section and the following one are largely borrowed from Pierre Riboulet, the great "humanist" architect and theorist of the 20th century, defender of the "legitimacy of forms" (Riboulet 1998). Riboulet approaches the question of the city through its modes of production. This approach based on the modes of production allows for the accumulation of a real genealogy on the composition of urban space, which helps facilitate a better understanding.

In order to understand the city, we must constantly bear in mind that any new division of space is inscribed in a previous division (see Figure 1.1), by replacing, modifying or juxtaposing it.

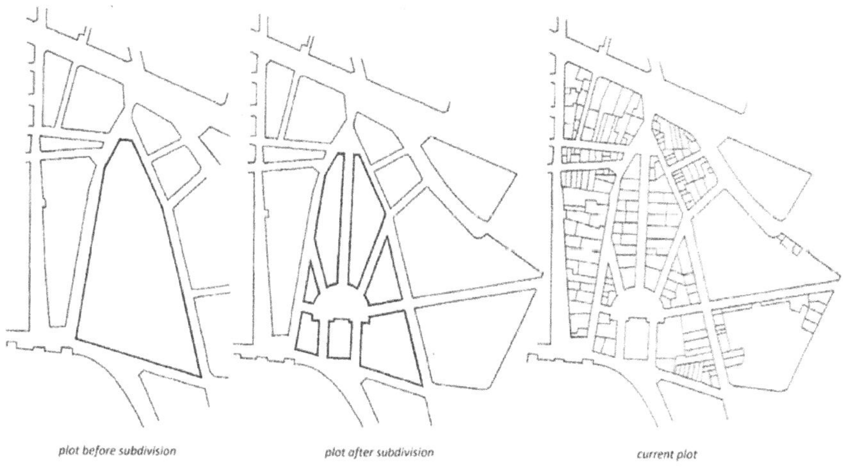

Figure 1.1. *Plot division*
(source: Riboulet (1998, p. 39))

Indeed, the city can be analyzed as an accumulation of fragments deposited by successive waves, with their forms and criteria contextualized spatially, technically and culturally, forming a resulting whole, the understanding thereof requiring a knowledge of successive "sedimentations".

The city must be understood as a process. It is the result of the continued growth of an original core that develops through free and available areas. Simultaneously with this development, a set of reconstruction and reconversion actions take place, with a corollary change in form and use. The urban evolution can be analyzed through the inertia of the weight of the past. There is obviously a time lag between the new social needs and the morphological and technical responses that meet them. Riboulet justifies the necessity of an urban composition imposed by the dialectic between this movement and inertia.

Knowledge of form cannot be attained from itself because it is the result of the social factors of production that determine it. The mode of production, of urban composition, shows how the city is made and how the factors of this production are combined. These determinants are essentially related to the level of historical, economic, social and technical development, the form of the social division of labor, and finally the cultural development of the different social classes and the mechanisms of ideological domination.

The effects can be grouped into three orders:

– production of specific urban forms;

– production of uses of space and the city;

– production of meanings.

Uses adapt to the initial spatial forms, and in turn, these forms produce uses. The same goes for meanings.

In order to analyze the city, Riboulet distinguishes four main modes of composition: the traditional mode, the princely mode, the liberal mode and the regulatory mode.

The *princely mode* (see Figure 1.2) is based on an authoritative relationship of intellectualism over practicality, in confrontation with the dominant culture. The archetype of this mode is the royal prerogative. Morphologically, the remarkable aspect of this mode is the brutality opposing it when contrasted to the ordinary space that surrounds it. These compositions retain a certain consistency in their production throughout historical periods. The princely space is a true laboratory of artistic creation that is in opposition to popular creations. This space focuses on outward representation. Today, this mode is only created at a reduced scale, the project of a mayor, a community councilor or a company CEO acting in an authoritarian manner.

The *traditional mode* (see Figure 1.3) is popular and vernacular. Cities of this mode are empirical; they have a system of streets, squares, a wide variety of buildings and conformity of the building materials used. We observe a typology of the constituent elements, but no identical repetition because any standardization, an outcome of the industrial period yet to come, is non-existent. These cities do not resemble each other because they are each the result of local adaptations of artisanal modes of production.

Figure 1.2. *The princely mode (source: Riboulet (1998, p. 78))*

Figure 1.3. *The traditional mode (source: Riboulet (1998, p. 88))*

The remarkable fact is the permanence of their characteristics; they impose their own conformity. There is strong consistency despite the absence of a plan or formal organization. This space is that of artisanal and individual practices, and there is no separation between the construction of buildings and their use. The cities that characterize this mode are those that assert their personality beyond the state[1]. The work of production is not isolated from that of creation, and there is no separation between the creator and the builder, between the builder and the user, and between form and use.

The *liberal mode* characterizes the capitalist era, the main determinant of which is the search for profit. Intellectual and manual work are completely separated. Here, we can observe:

– a brutal urbanization that has lost cultural references;

– the standardization of building elements;

– the specialization of trades and the separation between the production and use of the constructed space.

This mode of production also induces, as a corollary, a strong and speculative relationship maintained between land and the rent that it generates.

The urban space produced by this mode is characterized by the individuality that it symbolizes, with weak coherence and the fragmentation

1. The concept of the city-within-state versus the city-state will be defined in section 1.4.

of the signifiers. For its representation, it is enough to have the image of all the entry and exit points of our cities in mind.

Lastly, the *regulatory mode* is particularly essential for the other modes. It consists of numerous regulations, such as land use, and contents, such as hygiene services. This is the city-within-state mode, that is, the city-object. The state is content to control and limit the rights of private interests, facilitating private enterprise in urban matters by giving it a framework and rules. Here, the status of the mediator appears, indispensable since the regulatory city is built in spaces and times that are not known by those who enact the rules. These state personnel act in an abstract way, according to their own ideas of the city. The main effect of this mode is the standardization of production and consumption behaviors and the homogenization of space with the disappearance of eccentric local quirks. The effects on the urban form are very important, because they determine the constructability of spaces, densities, heights and the nature of occupation, which in turn has the consequence of determining their physiognomy.

1.2.1. *Combining different modes*

In real-world cities, these four modes coexist; there is not only juxtaposition but a real articulation between these modes. There is no, or very rarely, a pure mode. There is juxtaposition insofar as the city is an assembly of fragments produced by successive modes. There is a superposition of modes when we replace another. The combination can be genealogical. For example, the traditional mode contains a liberal component in terms of the right to property. The princely mode contains a component of the traditional mode with regard to the artisanal know-how of craftsmen. The liberal component, unique in absolute terms, can include a measure of princely composition, which has been calculated according to the development of land capital, or, for example, the display of a brand and therefore to the concentration of capital. The regulatory mode contains an element of the princely mode since it attempts to express power or arises from the expression of it. The liberal and regulatory modes are complementary because they cannot be sufficient on their own. In the liberal composition, the regulatory component may be more or less strong, for example, depending on the importance of a regulation specific to the area being developed. This regulatory mode needs the liberal mode since it cannot produce the city on its own, because it requires a significant concentration of capital.

The combination of modes is pragmatic. For example, during the construction of a city extension, we can observe a combination of different modes, except for the traditional one, which would have disappeared. From the princely mode, the personal intervention of a state official in the role of architects, town planners, etc., remains. The liberal mode is still present because of the value of the land, the presence of private capital and the need for profitability. The regulatory mode is omnipresent because the state has always sought to both control production and serve the other modes. Currently, composition as described by Riboulet's terms remains relevant. Any extension of a town or village originates from an urban planning decision resulting from regulations.

1.2.2. *Urban sprawl and morphological changes*

Over the centuries, and in general, cities have seen their populations grow progressively, despite some occasional declines. During periods when insecurity reigned, the enclosure, whose cost was important, prevented expansion and caused densification. When the enclosure became useless, it was the very functions of the city, their concentration combined with the demographic explosion, which made the phenomenon of densification persist. The extension was still limited by the problems of the means of communication and barriers such as the rivers, or bypasses which caused the same phenomenon of densification as the enclosure. In parallel to this phenomenon, the behavior of the population and its new needs led to the extension of the city in the form of suburban housing from the end of the 19th century.

This was the beginning of "urban sprawl", that is, the spatial expansion of cities, with the effect of peri-urbanization, discontinuity of the urban fabric and the standardization of living environments: natural and agricultural spaces were included in what Dubois-Taine and Chalas (1997) call the "emerging city". Today, sprawl is becoming synonymous with dysfunction. In this regard, Paquot (2020), while attacking the gigantism of current metropolises, denounces the problems caused by their size: multiplication of travel, increase in pollution, waste of agricultural land, overconsumption of energy, increase in environmental costs, etc. The solution lies in a change of paradigm: finally, today, it is no longer a question of relying on economic attractiveness to think about regional development; cities must be rethought in terms of the "harmonious living" of their inhabitants.

1.2.2.1. *Economic dynamics, social flows and impacts on construction*

In our market logic, housing and office buildings are durable goods, but perishable and tradable. These constructions are thus comparable to any other product. The duration of the life cycle is variable according to their own qualities, the desired aesthetic and the environmental conditions. The aim of the actors is to solve the problems facing the evolution of the city; however, the pursuit of profit tends to generate dysfunctions. There are at least two factors that determine the degradation of a neighborhood. The first is the evolution of needs, such as the rejection of collective housing in the 1970s, which led to the disaffection of this type of housing in favor of suburban housing. The second is the tendency to save on maintenance, the only way to fight against the attacks of time and therefore the degradation and devaluation of buildings. This process of housing devaluation and thus of district devaluation is the result of changes in population. The arrival of one or more tenants from a lower social category into aging buildings leads to the departure of the neighborhood in a more or less short term. This departure encourages the arrival of a category that is in turn socially less elevated. From one step to the next, the whole building suffers degradation and encourages a reduction in maintenance. This mechanism is identical to that of the neighborhood. The buildings themselves deteriorate from one to the next. The initial desire for a mixed society in the production of collective housing will not change this mechanism.

At the beginning of the 20th century, France embarked on a housing policy from which it gradually withdrew from the 1970s. In addition to improving the housing stock, the objective of this policy was to keep different categories of population within the same neighborhood. However, here too, we can observe numerous dysfunctions and the appearance of ghettos. Indeed, the production of rehabilitated housing has never caught up with the new demand linked to the evolution of family profiles and demographics, and these two principles amplify the deterioration. This mechanism has pushed disadvantaged populations into other, less expensive neighborhoods, which are generally more degraded. In general, the absence of intervention in the city entails the risks of an immobilist policy, that is, a sclerotic urban system that cannot respond to new needs. This attitude can lead a district to its ruin and even to its disappearance.

1.2.2.2. *Different actions on the urban area*

Society has different tools at its disposal with which to react to risks. It is the state that is responsible for taking measures to compensate for the inconsistencies of the liberal market. The state provides the legal means to intervene. Bulldozer operations consist of selecting degraded neighborhoods and razing them to the ground in order to rebuild new ones. First, a new system will be adopted. The evolution of needs provokes interventions on the very structure of the urban system. These interventions range from simple breakthroughs to the global reorganization of circulations, such as those of Hausmann in the 19th century. The consequences are different depending on the importance of the decision. The breakthrough can be an artery along which facades are built respecting the alignment defined by a prospect. In this case, the plot of land, which of course will not have changed, will no longer be orthogonal to the road. The most important consequence is that the intervention can lead to the total reconstruction of the plots. In this case, the new plot is reconciled with the old one remaining in contact to recreate the homogeneity of the structure. The main tool for these operations is, in France, the law of expropriation established in 1840. When the intervention concerns a complete block, whether it is considered to be insalubrious or never built on, the difficulty lies in integrating it into the existing structure. The operation consists of defining the complete system by first establishing the street circulation and then the plot.

The integration of the block with the rest of the site must be achieved at the architectural and social level. However, when buildings are designed for their own sake, without taking "into account" their relationship with the environment, the functioning of the whole fails. This is the major risk of neighborhoods created today, especially for so-called "sustainable" neighborhoods that "turn their backs" on that which already exists, on the pretext that they are innovative and the only ones capable of meeting current needs.

1.2.2.3. *Renovation*

When buildings are too run-down, there is a risk that the neighborhood will become ghettoized. To combat this deterioration, which has significant social consequences, local authorities can initiate or undertake renovation work. This type of intervention can be carried out on a national scale by assisting in the renovation of individual dwellings or on a neighborhood scale where the national aid system is reinforced. These actions have many advantages. By keeping the old structures, the development integrates into the site, unifying it, and maintains the morphological continuity in time and space. It strengthens the heritage of the building. Thanks to this coherence, we protect ourselves from urban and social functioning failures. One of the objectives is the maintenance of the populations, allowing the conservation of diversity and increasing the chances of an identical functioning. On the other hand, the operation is carried out, at least in theory, at a lower cost since it is limited by energy, hygiene and comfort standards.

However, these great qualities are only theoretical. First of all, it is not always possible or effective to maintain the populations that existed before the renovation. Moreover, the operation can be transformed into a segregation tool by rejecting the least well-off populations. Finally, insofar as it does not call these structures into question, it does not meet the new imperatives necessary for its evolution, as described above. Finally, the objective of social maintenance in the medium and long terms must be confronted with the social difficulty experienced during the execution of the work.

1.2.2.4. *Pattern*

In order to revitalize a neighborhood, the local community can initiate developments. It will be a question of carrying out a project whose impact will have the function of introducing the dynamics necessary for the renewal of the district in which it is located. This project is generally organized by a municipality or community, while the consequences are a function of dynamism and private demand.

1.2.2.5. *City extensions*

The extensions of the city have the same characteristics as the "bulldozer" intervention on the block, with the exception of those that integrate into the site, which does not address the same problems. Indeed, because the site is virgin, the link to be realized with the existing seems simpler to treat. However, these extensions require very particular attention with regard to their insertion and the projected management of their evolution. On the other hand, not being at the heart of the existing functions of the city makes its operation all the more tentative.

Finally, the extension of the city leads to the loss of its limits and its identity in relation to the countryside.

1.3. Difficulties of apprehension

As Mangin puts it:

> Cities have almost always been new at first, and most often in the form of a regular subdivision. Very early on, they had suburbs, initially called *faubourg*, which they gradually absorbed and reunited with the initial center. If their composition today presents complex arrangements, imbrications, irregularities, these are generally the result of the contribution of generations rather than of an *a priori* will. This reflection situates the importance of the "temporal" factor for the process of the urban fabric (Mangin (1999, p. 11), author's translation).

An action on the urban fabric therefore takes place at a given moment in its history. All the participants agree on the importance of the role of history and the need to clearly situate the intervention in relation to it. The urban culture acquired within this framework can teach us what can remain unchanged and allow us to inscribe the action without any contradictions.

The city transforms itself at all times; it develops and changes its meanings. This spontaneous and disorderly movement explains the need to take into account the time factor if we want to obtain a coherent analysis and urban perspective.

A new notion of time intervenes in the perception and understanding of the urban problem. It concerns the speed of communication and relativeness, which produces another level of complexity to the analysis: on the one hand, it is necessary to know how to distinguish the permanent from the ephemeral. On the other hand, it is pretentious to believe that we can recreate "ex nihilo" the heterogeneity of an urban space. It is thus necessary to question the city as both an object and a subject at the same time.

1.4. The city: object or subject?

In order to answer this question, we synthesize the reflections of Riboulet. The city is both an object and a subject: an object, because it is the product of a culture and a society at a given moment in history; and a subject that produces actions, behaviors and meanings. Its essence lies in the permanent relationship between the social forces that produce and occupy it.

When we study the city-state, there is a real identification of the inhabitants with their city. On the other hand, when the city is held by the state, by the organization of a centralized power, the personality is erased and there is a loss of a recognizable identity.

Moreover, the city's identity problems have long been associated with the poor quality of the buildings and the poor organization of the networks that generate the heterogeneity of flows. The actors recognize the inadequacy of actions that only affect the form of the buildings themselves, without questioning their relationships, or those that only manage the flows, but they agree that the real problems concern questions of urban form, the relationships that develop there and the associated morphology. The whole issue goes beyond the strict domain of technicians. Urban planning, which claims to be the "science of the city", is well founded on technical grounds. However, history always demonstrates the primacy of the social vision over technical development. There is no technical answer to a social question, and all attempts appear to be in vain. From the very first year of the construction of a new city, the social aspect takes precedence.

The necessity of composition comes from the dialectic between the movement of extension, of destruction–reconstruction, and the very inertia of the development, which brings a shift in time with regard to the social, economic and technical changes which command them.

The question of the form once evoked under the influence of materialist–dialectic criticisms, which assimilates it to a simple reflection or structural product, reappears in the field of the reflection and the action on the space.

1.5. The shapes and form of the city

The term "urban forms", in the plural, designates the spatial conformations of the city, of which morphological analysis has identified the elements (plot, building, block, street, square, etc.) and their system of articulation according to layouts and relationships, regulated by the mode of production specific to its social and historical context.

The term "urban form" in the singular covers a syncretic image of the historical European city (i.e., depending on the authors, either pre-modern-contemporary or pre-industrial). This image has a cultural and social value in its easy legibility for all. It is a civilizational asset, and for some, an "image of the collective unconscious" (Genestier 1988).

We can imagine that the key to the success of interventions on the city lies within this double relationship to "form".

Several decades ago, Mangin and Panerai (1999) laid down the framework for this problem very well. They describe the difficulty of responding to the growing need for housing in an expanding city without creating new public spaces. These new spaces do not allow for individual or collective appropriation. Since then, the problem of urban sprawl has increased, with those facing sustainable development only adding to it.

Finally, urban form can be approached as a representation. In the eyes of Joly (1988), the notion of urban form seems inseparable from that of the representation of the city. Joly shows how the study of urban form in history can be considered through the idea that we have of the city and its representations, and not just through historical knowledge of the urban fact. The idea is to look at these representations of the city as a chronological whole that provides a "process of knowledge", as expressed by Bachelard (1966). Joly goes on to propose that we examine whether there are moments and thresholds that mark stages in the understanding of the phenomenon or whether this only takes place in the ambition of the project (Joly 1988).

The debate on the "form of the city and the forms in the city" is old. This debate must be accessible to all if there is to be any consensus before political decisions are enacted. The stake is the realization of architectural projects in accordance with the city and not with the opportunity of the parcel of land, which generates rifts.

1.6. Representations of the city: "drawings and designs"

1.6.1. *Drawings*

A plan drawing is the "flat" representation of a building, city or part of a city, but it is also a program of coordinated actions over time. Until very recently, plans were reserved for spatial professionals and politicians. The introduction of the third dimension into these representations allows for the understanding of the plan through perspective, making forms and volumes more intelligible.

A brief historical overview shows that since antiquity, there have been two main families of documents. These are the more or less faithful drawings, whose use is clear. Between 3,000 and 1,500 BCE, cities were already being represented on clay tablets, including written reports and plans (see Figure 1.4) and signs which are rather symbolic, as well as very abstract images.

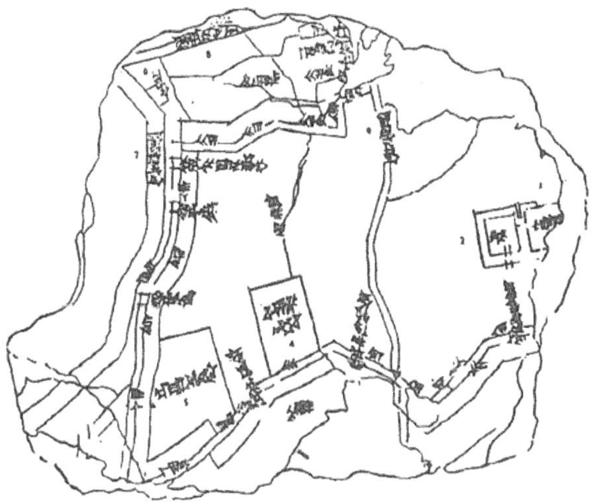

Figure 1.4. *Sumerian tablet representing the plan of Nippur city circa 1,500 BCE. (source: Benevolo (1995, p. 29))*

During the High Middle Ages, representations made use of iconography, which generated representations that were disproportionate in scale. Subsequently, with the Late Middle Ages and the Renaissance, these drawings improved in terms of the skill of the line, as well as in their geometrical competence. The technique used is a combination of the plane and a kind of perspectival folding (see Figure 1.5).

With the invention of perspective in the 15th century, we can see a real change in representation, since the same instrument seems capable of defining both the structure of the city and its spatial organization. The perspective view progresses and achieves an identity that meets between the form and its representation. The urban perspective becomes the "optical view". During this period, artists were initially asked to finalize the work undertaken at the end of the Middle Ages. The relationship between artists and their patrons evolved towards greater autonomy vis-à-vis the medieval corporations. Philippe Brunelleschi's work allows for a dissociation of the conception phase from the realization phase. This results in a more in-depth work on the characteristics of architectural works and urban planning. The representations are plans, elevations, as well as drawings aggregating perspectives and axonometries.

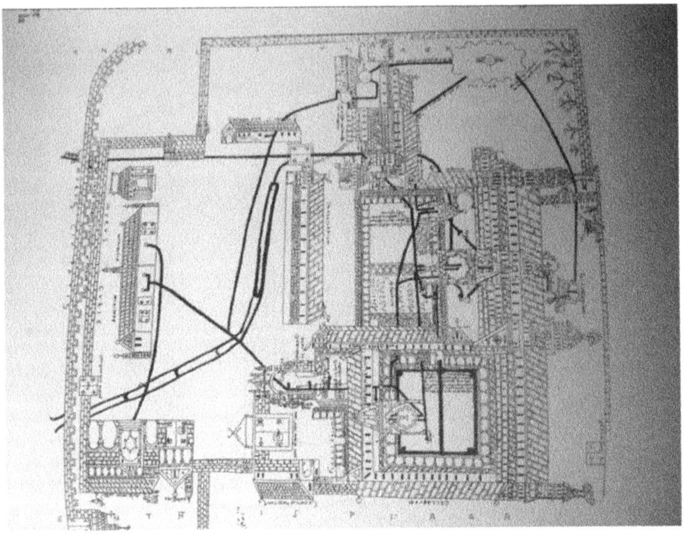

Figure 1.5. *12th century drawing; representation combining plan and drawdown, Canterbury Convent (source: Benevolo (1995, p. 186))*

Following the economic crisis of the Renaissance, the Baroque movement established itself in the 17th century. Artistic and architectural productions tended towards greater contestation and marking of the space with respect to the surrounding environment. Perspective began to be refined.

Towards the end of the 18th century and into the 19th century, the peak of the industrial revolution and the associated revolution in production facilitated the construction of long spans between load-bearing walls, and the standardization and fall in prices of expensive elements such as ironwork. Representations, especially volume, gain in precision.

Finally, in recent decades, we have taken a new step in representation with the computer-generated image. Digital tools make it possible to quickly represent a new project in perspective and give explanations over time. The drawing retains the performance of the perspective, but considerably increases the definition of the image to the level of "virtual reality", mobilizing the greatest sensitivity.

More recently, new methods for structuring and exchanging information around the project have been put in place thanks to building information modeling (BIM). BIM (both a digital model and a process for its realization) makes it possible to design, represent and visualize the project, in a collaborative manner between the different stakeholders. The images are increasingly realistic and dynamic. In addition to the architectural plans, execution plans and detail plans facilitate exchanges between the various trade-offs that occur at different stages of the project: from the design phase to the construction phase, the maintenance phase, and even its deconstruction. The same reasoning applies at the city level. The question is no longer that of the production of images and their modes of representation, but rather that of their circulation and reception.

1.6.2. *Designs*

The link between the design that presents the concept (the idea, the intention to achieve something, the project) and the drawing thereof (the transcription and representation of the idea) is a strong one.

In effect, drawing is a language. It makes it possible to express the intention of the designer or owner of the project. The drawing is therefore never trivial. It must make it possible to perceive and interpret the objective of the client.

As Pinson expresses it, "Drawing and design, though of the same etymological origin, distinguish, in French culture, form and substance, letter and spirit, where one would like, not to confuse them, but to associate them" (Pinson 2014, author's translation).

In order to better understand the link between the two terms (design and drawing), we rely on the definition given by DICOBAT (De Vigan and De Vigan 2014), according to which design is the act of creating and shaping a project and its plans. It is therefore, in the first place, an intellectual process, a rationalized creative thought, which calls for the observation of reality and its constraints, for translation of the final user/s needs, alongside the aesthetic sensibility of the designer. The thought is thus externalized and translated into a language (verbal, written, graphic) that can be communicated.

As an example, the study of "garden cities" plans, whether English or French, is a good exercise to compare a design (*dessein*) with a drawing (*dessin*) and vice versa. Indeed, the industrial, economic and social objectives are quite easily accessible from the reading of plans.

All these elements refer, through their materiality, to a culture, a symbolism and practices. As we mentioned above, thinking about the architectural and urban project means thinking about urbanity, that is, the multiple dimensions of the city, in addition to its materiality. In a methodological concern, the following paragraphs respond to a logic of decomposition, on the material as well as the immaterial level, of the constitutive elements of the urbanity. In a reverse movement, the meaning of the city will emerge through their recomposition.

1.7. The constituent elements of the city

1.7.1. *Limits*

The city has always demarcated its space through limits. This is what Unwin and Chabard (1922) called the self-contained communities of cities. This demarcation from the rest of the territory was either the rampart, which cut it off from the countryside, or the city-suburb divide once the latter had extended past the rampart. The rampart wall had the double advantage of reinforcing the military power and allowing for an effective control of the urban population. It clearly and visibly separated the territory of the city from that of the surrounding countryside. The city has never been totally separated from the countryside, and even if the city dominated the countryside, cities always retained agricultural and industrial activities in the countryside, at least until the 18th century. Braudel (1979), the historian, classifies cities into three categories based on their boundaries. There are cities open to the countryside, such as Greek and Roman cities, cities closed in on themselves, as per the Middle Ages, and cities held by the state. Cities see their extensions managed differently according to their political organization. For example, Greek cities refused expansion by limiting populations to an ideal number of around 20,000 inhabitants (Braudel 1979).

For a long time, the limit of the city was demarcated by the rampart which initially had a religious meaning, followed by a military one. Sometimes, it is an elevated site that acts as a rampart or reinforces it. The boundary may also be of a different nature; for instance, it may be a canal, a railway line or a highway. It can also be natural, for example, a river, a mountain or the sea. The limit has the characteristic of delaying the moment that a city develops, and for this reason, participates in the phenomenon of densification.

There is, historically, a loss in differentiation between the city and the countryside because the city encompasses the enclosure in its own form, abandoning its previous form, characterized architecturally by the presence of the wall. At this point, there is no longer an inside and an outside, but instead a center and a periphery in the modern sense.

In the first part of the 20th century, up until the 1970s, we can observe a dilution of this limit by the significant growth of the suburbs, as well as through zoning. These spaces with specific functions cannot fully claim urban status, whose functions are varied and multiple and whose morphology is produced over time.

Today, the limits of space are constantly being pushed back, not only thanks to means of transportation, but also by the arrival of digital technology. The very notion of a limit is becoming obsolete, especially when we think of the city from the point of view of lived space, or even lived spaces (plural). The question of limits is now posed in terms of accessibility and temporalities: from a phenomenological point of view, it is the time factor that allows for the measurement of distances to be traveled, and it is still the time factor (synchronous time) that allows us to access increasingly distant spaces, thanks to digital technology. The dilation of a city's space-time impacts its structure.

1.7.2. *Structure*

The location of a city is dictated by a political decision justified by strategic or economic reasons. Thus, we can see that in the past, the selected location will be situated on higher ground, when insecurity reigns and when it requires a defensive position. On the other hand, when the period is less troubled, it is the economic and commercial functions that take precedence. The chosen location is therefore the one representing the best compromise between the means of communication, the possibilities for trade and the site's natural resources. The ideal site is one that easily combines accessible arable land, mineral resources, a means of communication such as a river or a sea and, finally, a central location between the greatest number of other cities. Once the site has been defined, the establishment will be carried out according to the best possible local conditions and in search of a microclimate.

Often, the role of climate, in particular sunshine, has been taken into account. At some sites, the sun's rays are absent all day and all year round, preventing warmth and causing feelings of discomfort. In addition, the problem of sunlight arises with regard to the orientation of the buildings themselves. In the northern hemisphere, for example, the trajectory of the sun prevents the north-facing facade from absorbing rays, with the exception of summer, when it does so for a short time at the start and end of each day. The nature of this phenomenon not only requires reflection on the orientation of the building's openings and organization of rooms within, but also that of the strip that is formed by adjacent constructions. When constructions are built in rows, a characteristic organization of the city, only two facades are open to the outside. When the facades are oriented North–South, the first never receives sunlight, while the second benefits from it all day. An East–West orientation of the facades is therefore preferable since it allows each of them to receive a quantity of lighting and energy for half the day. This last orientation is therefore the best suited to this type of building organization and as such, a fortiori, if the width of the building is a factor and two non-traversable dwellings are built in the space, then owners, whose homes are located north of the new construction, can pursue legal recourse and obtain compensation.

One particularly pertinent question on the topic is yet to be studied. How will global warming affect the status quo? Between the search for comfort through cooling and the need to receive the sun's rays for health and mental balance, an arbitration may become necessary.

As far as the wind is concerned, the consideration on this climatic element goes back a long time in history, and although it cannot be accurately dated, Vitruvius established the importance of the phenomenon in his treatise "*De Architectura*". Therefore, the site, chosen for its location and resources, must specifically take this element into account. For example, habitations often choose to shelter behind a hill for protection from a prevailing wind. Hence, the very organization of streets also plays a role in preventing winds from rushing through them. Public squares, for their part, have dimensions and heights constructed so as to shelter people from the wind, thereby allowing for a pleasant experience of the space.

The Roman grid was orthogonal and oriented North–South and East–West for religious and symbolic reasons. However, this orientation was adjusted according to the wind directions of the site, applying a slight rotation of the grid if necessary. This organization is schematized in Figure 1.6.

Vitruvius, in his work "*De Architectura*"[2], already specified that it was necessary to prevent regular winds from passing directly through the streets because, cold or hot, they were always deemed harmful to health.

In addition, Rossi (1966) considered the map to be a "singular" element, in the same way as a temple or fortress, and therefore an element capable of accelerating the urbanization process of the city.

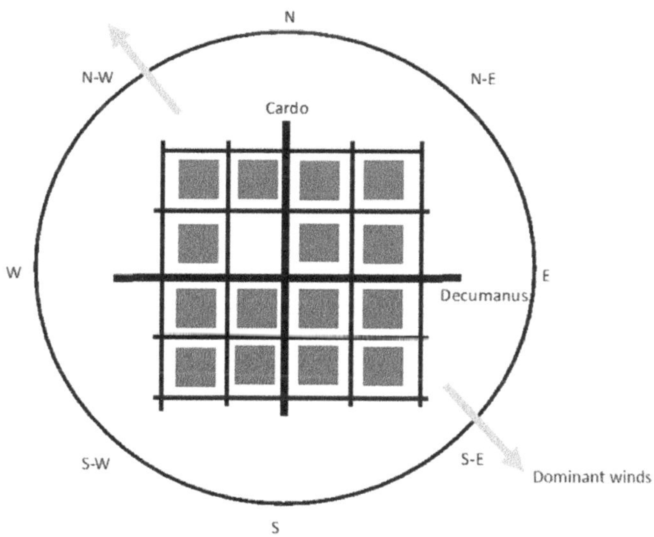

Figure 1.6. *The establishment of a new Roman town (source: author). For a color version of this figure, see www.iste.co.uk/adolphe/integrated.zip*

2. "*De Architectura*" is a treatise in Latin, written by Vitruvius around 15 BCE, and composed of 10 books that have been translated into several languages throughout the course of history.

Regarding the map, Lavedan specifies that whether it is a "spontaneous" city or not, the pattern of the map is never due to chance (Lavedan and Huguenev 1952).

Mumford (2011) contradicts the interpretation that the geometric map developed in the Middle Ages illustrates a transition from a dynamic to a sclerotic culture.

Finally, in an era of globalization, multiculturalism and sustainability, what should the structure of the 21st century city and the city of the future be? A city that must reconcile the ecological long term and the necessarily rapid transformations of the urban system, a city that must also reconcile the various uses of its spaces. Of course, the transformation of cities from places of production of material goods to places of production of information has had an impact on the organization of urban and peri-urban spaces: spatial fragmentation is also social. Once again, it is about accessibility, this time equipped with the knowledge to ensure social inclusion.

1.7.3. *Layout*

The study of the pattern, as a constitutive element of the city, is neither geographical nor sociological, but instead morphological. The gaze is therefore that of the architect and the elements are analyzed, according to Panerai et al. (1980), as urban elements.

By pattern, we mean the marking on the ground, that is, the footprint of the city. Its importance is sometimes downplayed, yet this abstract element has very real consequences for the city and the life that unfolds there. We could even say that the original pattern is the main determinant of the morphological evolution of the city through the phenomenon of permanence, as evoked by Rossi (1966) and Poëte (1929). The phenomenon is simple because it underlines the lasting nature, sometimes on the scale of the millennium, of laying down the groundwork for buildings that will succeed one another over time.

The study of the pattern provides a better understanding on the development of almost all of the major European cities. In general, we can correlate the pattern of cities to a limited number of generative determinants. This concerns the specific character of the site, the agricultural measurement which gives its dimensions, the voluntary or involuntary establishment of a landscape and the renewal of existing roads.

We have essentially inherited two types of patterns from antiquity and the Middle Ages: the checkerboard and radio-concentric forms, to which we can add a third type: the linear pattern (see Figure 1.7).

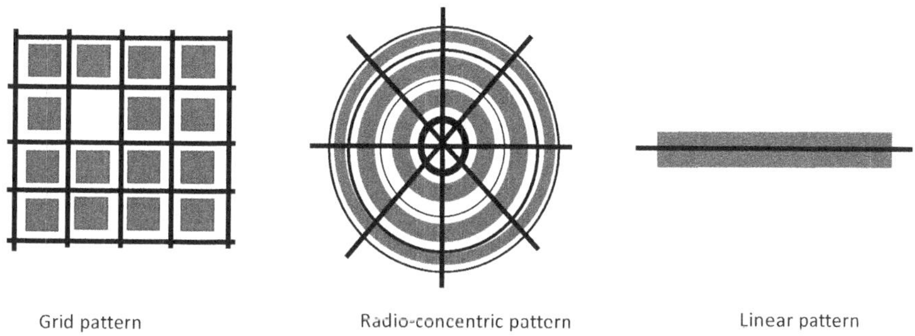

Figure 1.7. *The typologies of city patterns (source: author). For a color version of this figure, see www.iste.co.uk/adolphe/integrated.zip*

The Etruscans used the grid pattern almost 1,000 years BCE. The foundations of the city had a religious character, and the grid was a symbol of the cosmic order. For this reason, its orientation was that of the cardinal points: the North–South (*Cardo*) and East–West (*Decumanus*) were symbolic delineations set out by priests. The Greeks adopted it later by Hyppodamos de Miletus, who used the pattern to exploit a close concordance between the material structures of the city and its social order. The Roman organization is inherited from these Etruscan and Hellenistic cities. The characteristics of the grid facilitated the establishment of new cities across the empire. The loss of the religious referent allowed for a more rigorous adaptation to the wind, as advocated by Vitruvius. There followed a checkered division of the spaces.

The checkerboard grid pattern, based on the orthogonal grid, was the most effective, least tendentious and fairest means of spatial division, in a colonialist approach to space. The organization of cities in a "checkered" pattern is characteristic of a pre-established organization, authorized or organized through the decisions made by a sanctioned power. It is the most practical and can be found at the origin of a great many European cities. Although predefined in its dimensions and orientation, it remains flexible in its application and generally adapts to the landscape. The regular arrangement, also known as a "chessboard grid", in which all the streets are parallel and perpendicular to each other, intersecting at right angles, is technically and socially the easiest organization. Such a system normally involves a square or rectangular city. The term chessboard should not be taken in the strict sense. In an actual chessboard, all squares are square. To define the main characteristic of this type of pattern, the following definition is useful. We also speak of a grid, a frame or a reticular grid. By reviewing Lavedan's 1959 work "Geography of cities" in detail, it is necessary to distinguish "chessboard" or "checkerboard" pattern, where the blocks are squares and the orthogonal pattern, where the rectangular blocks can have edges of variable dimensions. We have observed this type of pattern throughout history, in the cities of the Roman empire, then the medieval fortified bastides and even in the cities of the New World.

The checkerboard pattern was not easily applied to sites with irregular topography, but it made it possible to divide the entire territory in a simple, fast and above all equitable way. This made it possible to avoid conflict. In addition, this division avoids residual spaces, makes provision for future extensions by eliminating any degradation of that which exists and helps mitigate the disorientation experience of newcomers. This pattern reinstates the street as one of the primary elements, participating in the whole, in addition to the narrow passages that still endure. The existence of the streets is therefore recognized.

This checkerboard pattern lasted until the early Middle Ages. During the first part of the Middle Ages, pandemics, famines and wars caused the population to withdraw into itself. We have to imagine, for example, that the population of certain cities, especially those of a medium size, would return to confine themselves entirely within its auditorium. The rest of the city would be left abandoned, unprotected and the victim of looting. It was not until the late Middle Ages that we began to observe the commercial explosion, which encouraged the development of transport structures favoring trade through radial traffic networks. From the late Middle Ages, the old towns began to look inwards, their pattern adopting a radial structure, which in turn saw the closure of many corner buildings. Then, as of the 13th century, the new towns resumed using the orthogonal grid, mainly for its practicality in allowing the lord in power to manage and exploit these spaces, even if the peasantry working the lands were lawfully free.

The *radio-concentric pattern* is observed when the city itself is the crossroads for connections and the seat of interests, its growth is star-like. This pattern was also found in towns of military and defensive origin as of the 18th century. This urban form typically results from the significant growth of a city high street, whose proximal appeal radiates outwards. Generally, the pattern respects the topography of the site. In any case, its root cause for its evolution is due to its influential connections with other cities. It likewise can be the result of an erosive-extension process of the checkerboard pattern mentioned above.

The big cities of the New World, particularly those in Australia and South America, are no exception to this pattern. However, these more recent cities did not experience the dark period of the High Middle Ages and its phenomenon of contraction. Therefore, they always have this original checkerboard pattern, adapted to the local geography of the site.

In making a comparison of the checkerboard and radio-concentric system, the sociologist and anthropologist, Hall (1966) describes the checkerboard system as "sociofugal", for its tendency to dissolve functions and activities. He opposed it to the radio-concentric system, which he describes as "sociopetal", which to his mind concentrates them. The first consideration regards pedestrian orientation, whereby an error outside the 90° right angle becomes more noticeable in the checkerboard pattern than in the radio-concentric system, as the error will be low at the start, but will get worse as it progresses outwards. For some people such as Sitte (1996), the checkerboard pattern is criticized as being a purely administrative production, which cannot create enthusiasm since it prevents any room for surprise and the variety of perceptions otherwise offered by winding streets and squares, as found in the cities of the Middle Ages.

In contrast, this system is praised by individuals such as Lewis, who describe the achievement of the pattern as efficient and defend it from all notions of monotony and ease. He specifies that the establishment of public spaces is not a matter of chance and is better designed so that these places participate in the balance and functioning of the city. Of course, we can relate these views to the original culture of their authors.

In addition to these two types of patterns, a third is added: the linear pattern, resulting from the growth of a hamlet or a village, which has a communication line running through it, connecting it to another town, and as such, developed along either side of it. This type of organization is called a "street-city" and is usually only found at the heart of small towns.

Urban grids did not undergo any essential changes until the industrial revolution. In fact, it was not until the 19th century that we observed the development of the grand boulevards and so-called Haussmannian innovations. Generally, these breakthroughs, recognizable thanks to the configuration of the facades, have the goal of improving circulation fluidity at their origin, as well as the better-known desire to facilitate the maintenance of social order. At the same time as these innovations, all the towns with ramparts and neighboring land free of any construction started developing circular boulevards. The demographic explosion of the "glorious thirties", the development of surrounding suburbs and the democratization process brought about by a surge in individual motor vehicles led to a last major development, namely, that of ring roads.

Today's cities are composed of fragments from different, juxtaposed, patterns, the result of their unique evolution, often difficult to comprehend as a whole. Their pattern is therefore described as one of these three types (checkerboard, radio-concentric, linear) or the composition of two and even sometimes all three of them:

> Today, everything suggests that the city of the future will have to combine the strengths of different modes of transport and for this, checkerboard town planning will need to articulate the design of road networks that facilitates the flow of traffic, linear town planning in connection with the development of individual sites dedicated to public surface transportation, and circular urban planning, around train stations and metro stations (Laterrasse 2016, author's translation).

Knowledge of urban patterns not only provides for a better understanding of the morphology of the city (its structure, the axes of its configuration, the location of built and non-built spaces, the gathering role of a monument, etc.), but is a necessary element for any practical intervention.

1.7.4. *Plot division and private property*

The notions of plot division and private property are important for the phenomenon of densification of the city. Intervention constraints for composition lead to zoning playing a template role during its evolution. Any new division is part of a previous division. The population explosion, the industrial revolution, the concentration of city functions and the action of social mechanisms lead to a significant need for land. This increase in demand produces an increase in costs, with prices graduating according to their location. Land is not like any other product. It is not the object of any wear and tear and is only the object of the needs to which it can or will be able to respond. Fujita (1989) describes it as a permanent revenue. It is therefore the object of speculation. An increase in costs and the desire to make more profit results in its division into smaller plots. It should be noted that the value of housing is not only linked to the housing structure, but to its location and proximity to the amenities that surround it. It should also be noted that this speculation is only valid in the context of a positive demographic pressure.

A particularly special trait for this type of product is that the type of plot created in turn becomes a decisive element in a city's morphology during its evolution since it essentially acts as an unavoidable production template for the city in question.

As the proceeding section outlines, the division and initial choice of the building composition has a significant impact during a period of densification. The alignment with the street facilitates this densification and its involvement with the city center where most commercial functions occur. For this functionality, the most suitable pattern is the corridor street, which channels traffic and frees up space at ground level. Other uses such as residential and office functions are transferred to the upper floors. The consideration to set back structures from the alignment helps avoid ventilation and sunlight issues, but in doing so, lowers the capacity for densification and is therefore only suitable in peripheral areas. The increase in the cost of land at the center shifts demand to the periphery, causing the latter to increase in cost. Overall, in a caricatural way, this phenomenon leads to the rejection of less well-off populations living outside of cities and causes the densification of the central zone in order to distribute the additional cost. The high cost of land can only be supported by increasing the number of users, that is, by increasing density. Hence, we observe a correlation between the height of a city and its land value. Throughout its evolution, a city will tend to do away with intermediate cases between public and private land uses. On the other hand, it will allow for a change of plot use and not impose any limitations. In terms of general organization, monuments, squares and streets are a part of what has been termed "hotlines". During its evolution, any distinct identity of the new organization may produce an architectural or functional inconsistency insofar as it ignores or poorly respects that which is pre-existing. Densification is first achieved by using any gaps or open areas that were left between buildings, within the property's perimeter or along the boundary. However, due to the high cost imposed by limitations, these spaces are few. Initially, sizable predominantly unbuilt plots are divided, thereby providing spaces for a number of new properties to be built. Once these gaps have been filled, the plot itself can be the site of densification. This can take place in a variety of different ways. The new constructions can be built at the bottom of the plot, or along its length, or on both at the same time. This typology of the built environment leads to the disappearance of open space, ventilation and sunlight accentuated by the reduction of the number of openings and the multiplication of daylight shafts. After maximizing the built area in relation to the plot size, densification is achieved by raising the building height, through the superimposition of additional floors, and finally by making the best use of the depth below the surface. The lack of parking space within these blocks results in a rejection of vehicles on the public thoroughfare. Parking spaces are therefore removed in order to return public space to the public, thereby reducing the number of intramural vehicles. When the need for densification becomes too great, this phenomenon is reversed and plots that are too small are grouped together in order to allow for the construction of large construction equipment. The physiognomy of the city is thus organized in a concentric manner. Buildings increase in height and density the closer we get to the city center. As we move away from the center, they become lower and more dispersed, leading to a sprawl in the residential areas. For several decades now, current urbanization policy has focused on methods with which to combat this urban sprawl.

1.7.5. *The city block*

The city block is a network of complex interpenetrations, fundamental because it guarantees the mixing of activities and uses. The result appears as a set of private plots unified by a network of roads. The constitution of the structure, in its simplest form, is a homogeneous allotment, and thus appears as the staggered superposition of two irregular grids. In the first, each grid is formed by a segment of streets with rows of constructed plots along them, while the second comprises the four streets that isolate the block. The block is not, however, a block that was laid down a priori, but rather arises from the addition of plots that are open to the outside, onto different streets which convene at the center along a common boundary. The block is thus, in the simplest case, formed by two rows of adjoining plots and its width is the sum of the depths of these plots. On the initial grid, the building changes without questioning the overall structure, ensuring the permanence of the pattern. The block is thus a constant of the overall morphology. The continuous line along the bottom of the plots divides the block into two distinct parts. This line is sometimes constituted by a service passage that allows for both rear access and the possibility for the removal of garbage and waste water. The dimensions of the blocks are dictated by a compromise between the practical necessity of passing from one street to another without any considerable detour and the most economical use of the lanes. These dimensions vary greatly from one city to another, ranging from 40 meters to more than 200 meters.

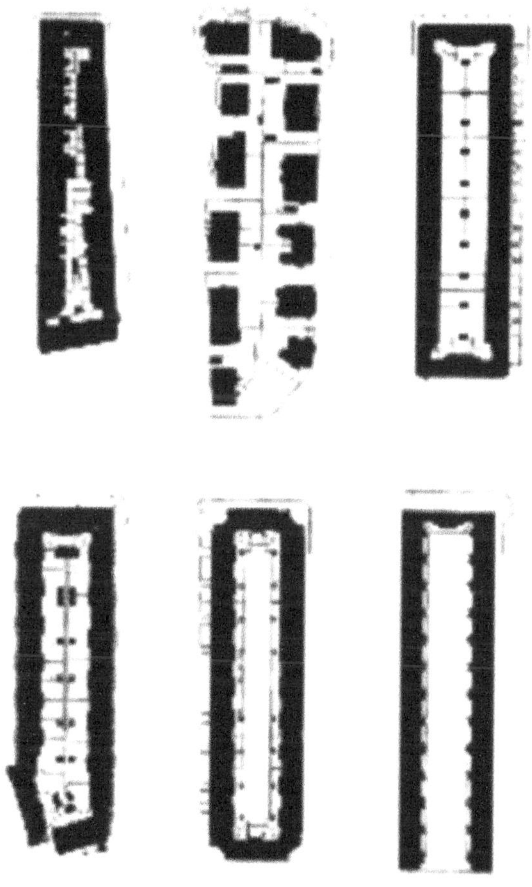

Figure 1.8. *Example of city block shapes (source: Panerai et al. (1997, p. 170))*

The grid having been defined in advance is the difference of the speed of the "filling" that gives rise to a greater or lesser variety in construction. The different faces of the block are not equivalent. They are the object of a hierarchy and are assigned various functions. While remaining in the scheme of a simple and regular whole, these buildings must allow for the operation of certain functions. For example, the plot may be wider to accommodate shops and the buildings may be set back on the first floor to allow for an arcade bay. When the block is rectangular, we can sometimes observe a return on the plot and an internal circulation in T.

Generally speaking, the faces of the blocks are a function of the hierarchy of the street to which they relate. The less accessible block cores are less constructed, occupied by a set of gardens, courtyards, utility buildings, lower and more recent than the buildings on streets. There are many different block shapes. Figure 1.9 shows some examples.

The city blocks and streets have evolved over time with a more or less important opening to the outside according to the risks, real or perceived, from the outside. During the end of the 20th century, the block began evolving towards a deconstruction. We can notice an inversion of those that are occupied and those that are vacant (see Figure 1.9).

Since the 1970s, peri-urbanization[3] has become the model for the development of new blocks called subdivisions. However, these subdivisions do not have the function of urban blocks, notably because of the absence of the street in the sense described above. This street is meaningless and only has a traffic flow function. In spite of the recurring incantations for the densification of city centers, presented as essential for the safeguarding of nature, peri-urban sprawl still seems to be in effect. This mode of development of peri-urban space is found even in "gated communities" models, which are very common in the United States, but not completely absent from European landscapes.

The change of typological scale that occurs when the unit of intervention is no longer the house or the built-up plot, but the block or subdivision, constitutes another type of rupture in urban production (Panerai 1988).

There is also a real loss of urbanity in the commuter towns or districts, distance between the workplace and place of residence, problems with transport and communications or lack of facilities (see Figure 1.10). This

3. Peri-urbanization is a complex planning concept that includes many profiles. This notion has been described through a social approach in the publications of Jaillet (2004).

rupture is not felt at first glance and is only later criticized, indirectly, by socially concerned professionals, such as sociologists and psychologists, regarding the everyday problems that this dynamic can produce.

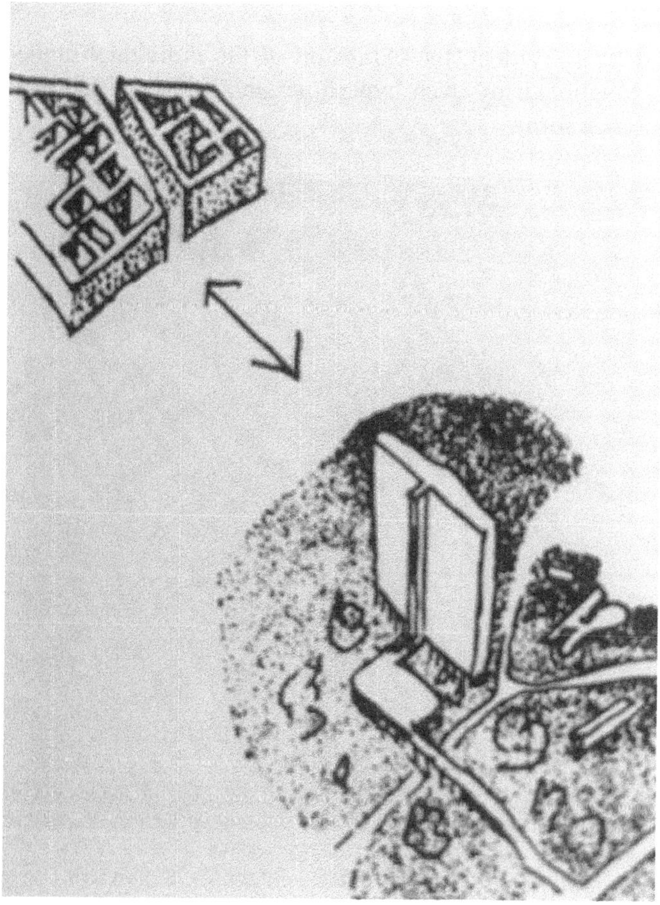

Figure 1.9. *The Parisian project for the city center of Paris, the Le Corbusier Sketches (source: Krier (1975, p. 75))*

In this context, the simplistic zoning functionalism executed by urban planning, which hierarchizes the roads and streamlines public transport, has shown its limitations, in particular, its inability to grasp urbanity. This complex phenomenon is established through a dialectical relationship between the city and the citizen.

As Krier (1975) expresses, within these projects, the optical fascination emanates from isolated architectural elements and their aesthetic perfection, and not from the notion of an urban space composed of geometrically definable squares and streets.

The modern movement, the explosion of the individual mode of travel, zoning and functionalism, even hyperfunctionalism, have led to the loss of the historic block form.

Figure 1.10. *The loss of urbanity profile, Slotervaart, Amsterdam (source: Benevolo (1995, p. 456))*

Today's city is made up of either blocks which have endured since antiquity or the Middle Ages, or city blocks resulting from the various transformations of the modern and contemporary eras: therefore, in addition to the closed Haussmann style blocks, there are the fragmented blocks located on the periphery of urban areas and the open blocks of denser urban areas. Following the study of the typology of blocks, in the 1980s, Christian de Portzamparc created the concept of the open block (see Figure 1.11), which, by separating the buildings (which are no longer adjoining), multiplies the perspectives and the propagation of light, through the innovation of staggered faces (Accorsi 2010).

Figure 1.11. *Drawings for the Masséna project, 1995 (source: Christian de Portzamparc Architecte/2Portzamparc)*

With the goal of a sustainable city, the city block must increasingly anticipate the needs of its own functions: the presence of green spaces, the management of rainwater and gray water, parking, its ecological impact, the integration of renewable energies, the pooling of equipment, the management of the different usage times, etc.

From an operations point of view, the characterization of blocks is both an administrative and a project tool. For example, it suffices to think that the morphological components of the block are at the basis of the formation of Urban Heat Islands (UHI)[4]. The definition of the reference entities for each block (density of the organization, age of the building, construction materials, exposure, greenery, energy requirements, etc.) becomes the key elements with which to anticipate the development of these urban fragments and better manage the phenomenon of climate vulnerability.

1.7.6. *The street*

Streets have a specific role and function in the city. The street is a public space of particular importance to the city. For this reason, it is dissociated from other public spaces. Pöete (1929) explains that, on the one hand, the city is born in a specific place, but it is the street that makes it live, and, on the other hand, it has the character of permanence in time. Rossi (1966) imbues the street with a special status by considering it as a "singular element", that is, an element capable of accelerating the urbanization process of the city in the same way as a monument. In the perception of the city, Lynch (1969) shows us that the street participates in its "imageability" and that it is a predominant element. The dimensions (height and width) of the facades, the frequentation, indicate to the passer-by the level of importance of the avenue. A trustful feeling will emerge when the identity of the street has visible continuity, and on the contrary, a street may elicit confusion when it lacks identity. Lynch notes that when the origins and destination are clear, the street has a strong identity. Roadways can also have stronger or weaker "directional" and calibration qualities.

Evidently, streets thus have variable characteristics and are hierarchical in their perception.

4. The term urban heat island refers to the artificial microclimate caused by the mineralization of the territory and human activities, producing an increase in temperature in the urban environment, thus contributing to global warming, with consequences both in terms of health and biodiversity.

Often, the street, and especially its sidewalks, are the access points for so-called "local" businesses. These businesses have a special status. They are semi-public and semi-private places. The merchant, in order to maintain the attractiveness of their store, maintains and monitors the access space, that is, the street and its sidewalks. They therefore have a social role in regulating and structuring the street and, by extension, the city. How do we perceive streets without commerce? Do pedestrians walk along them? What feeling of security or insecurity do they invoke? By extension, what shape would a city be if its only means of supplying society were peripheral or online supermarkets? Fortunately, at least in Europe, the trend seems to be heading in the opposite direction and we are seeing the proliferation of local distributors.

Each street has its own physiognomy. For this reason, Panerai (1988) explains the impossibility of defining ideal dimensions and generalities. The street is not an invariant object, and the constants that can be found there must be constantly replaced by the urban culture of the city where it is located. Defining the width of a road and thereby regulating its technical capacities for a certain period of time is only one aspect in the design of public spaces. Pertaining to the design of the roadways and sidewalks, the question arises of the more or less precise character of its boundaries on both sides. Indeed, the continuity of the road does not automatically lead to alignment and the semi-detached use of a building. These dimensions have often only been obtained as a result of densification following on from the original settlements. In order to regulate the relationship of the street to the built-up plots that border it therefore presupposes different strategies whose constraints can range from the simple definition of width to a pre-agreed obligation to conform to a given facade style (see the regulatory mode). The streets directly serve both sides of the built-up plots, while at the same time allowing for movement within the neighborhood. The street allows for movement and perception; it is the support of the building and in this dialectical relationship that we can observe the conditions for overcoming sectorial approaches that reduce the city to a collection of technical objects. The street is not only a space of circulation, but also the passage of sewers, gutters and various networks underneath the sidewalks. At least for the European case studies, when we study the evolution of the street and the technical equipment buried beneath it, we see the relative progressive sinking of the latter through the successive accumulation of rubble that is the result of interventions. Its functions have evolved over time with the evolution of uses, means of transport and chosen priorities. Therefore, at certain times, we will find no sidewalk, only to discover a step or two or sometimes several ahead, that there is traffic and parking bays, leading on to strictly pedestrian sections that totally exclude vehicles, or seek to contrive some sort of compromise. Outside the cities, developments will give rise to increasingly important roads, such as boulevards, avenues and then ring roads.

Intersections are generally the result of the pattern choice, unless they have been the object of a composition work, such as that of the 19th century that produced exchanges between public and private spaces.

Today, in city centers, we can observe the disappearance of the specialization of traffic. Often, the car is excluded, while the so-called multimodal means of circulation see an interpenetration with a quasi-mixture of pedestrians, cyclists, scooters, personal transporters, trams, etc.

We insist on the importance of the street's structuring role. It has a primordial sensory role insofar as it is the street that makes it possible to understand the city. Its perception can be more or less precise, depending on its composition and the culture of its user. Finally, it has a social role since it is the street that facilitates exchanges between public and private spaces.

Huet (1986) points out that for most people, the definition of public space is in opposition to private space. This simple definition, which refers to the use of the space and its legal character, is extremely insufficient insofar as it does not confer any precise formal quality to this space. This is the reason why, in the practice of modern urbanism as conceived by planning technicians and architects, this empty and hollow space has become a residual space. It is what remains between private spaces occupied by architectural objects that are often formally overdetermined. It is therefore necessarily a space without any specific form, without precise symbolism, without a name, insignificant and unnamable in the etymological sense of the term.

The modern movement has engendered the reversal of occupied and vacant, but above all, it has transformed the use of the street, with the appearance first of the tram, then the steam locomotives and, finally, the automobile and the construction of urban expressways achieved in the 30 glorious post-war years (*Les Trente Glorieuses*).

The contemporary era has seen a reversal of this trend, to the detriment of functionalism and in favor of development projects aimed at reducing automobile traffic within the city and developing alternative modes of transportation. This involves reclaiming these traffic spaces and transforming them so as to reconcile mobility, urban planning, lifestyles, services and sustainability.

Advances in the means of transportation have played a major role in the evolution of the city. Beginning with the use of the vehicle, which began as an animal-drawn vehicle and later became motorized. It facilitated the transport of people, food and materials. It has, therefore, enabled the expansion of the city. The attainment of increasingly sophisticated means of transport reduced travel duration. Subsequently, the densification of the population, which was to produce the saturation of the network, led to the implementation of public transport, in particular the development of underground transport. This last type of transport has the characteristic of making travel times more or less constant no matter the number of people traveling. There no longer seems to be any limit to the expansion of the city, the emulation of which is reinforced by the transport infrastructure on a national and international scale. Progress is not only limited to the physical transport of people, but also that of intangible communication, such as the telephone, radio, newspaper, cable, Internet, etc. Entire regions have lost their identity insofar as the materials used for construction were historically local, while new means of transport and their industrialization have attenuated the local renown of any achievements. Today, we are promised autonomous transport, which, on the one hand, should reduce the risk of traffic accidents, and, on the other hand, open up places that are not accessible via public transport. We can note that for a long time, people have sought ways to reduce travel time, which means reducing the physical distance. Today, this research seems less, with more obvious favor being focused on economic and ecological optimizations.

The physiognomy of the city in its verticality has long been limited by technology and materials. The pressure capacities allowed by concrete and steel in construction have allowed humans to construct higher and higher buildings. However, among the inventions of the 19th century, there is one invention in this regard that towers above the rest and revolutionized the physiognomy of cities worldwide: the elevator. This invention allowed for buildings of theoretically unlimited height.

We could speculate endlessly on the true motivations of people wishing to be closer to the sky, unless it is just the desire to colonize this space and demonstrate their power.

In general, horizontal and vertical travel assistance has practically nullified the limits of the city's ability to expand. However, this reduces the user's physical involvement and totally alters their kinesthetic perception. The physical effort to reach a place only provokes weak sensations. Places start to look more like an address in the computer's memory than the finality of a physical effort, induced by the succession of spaces.

Moreover, the production processes have changed with the times, from the one-off self-constructions until the end of the Paleolithic era, to the mass-produced uniform constructions of today. The construction elements have therefore been standardized. This type of production made it possible to reduce costs through the economy of scale and the experience curve. At the same time, they made it possible to meet the needs of a growing population. Originally, it was the functions that produced the characteristics of the manufactured elements. Later, the definition of standards was the result of a search to maximize the service rendered in relation to the costs. Today, it is the functions of ergonomics and aesthetics, associated with the previous types, that hold the attention of product designers. From a concept of regional aesthetics through artisanal production, we have moved to international design. The use of these elements and materials, with standard characteristics, further reduces the means of expression, slowing the march of creativity and its interpretation by its inhabitants.

Technical progress and knowledge about hygiene have improved well-being and comfort. The street has technically evolved a lot. Originally, it was made of compacted earth, with a central gutter and later lateral gutters to improve traffic safety. The needs and technical means having evolved greatly, the evacuation of rainwater and wastewater changed to an underground system of sewers, thus releasing the space for more or less easy circulation. The distribution of drinking water in each building and then in each household, electricity and the collection of household waste are the tools that have made it possible to mitigate unsanitary conditions in the city.

Hygiene and comfort require compliance with sanitary equipment, ventilation, lighting and a minimum surface area. All of these constraints have given rise to regulations and the development of prospects. Originally, cities were composed of one- and two-story buildings with a height ranging from 3 to 6 meters. Given the width of the streets, often more than 6 meters, the sun penetrated the street widely all year round. The facades that did not face north were therefore sunny for a good part of the day or even the whole day. The increase in building heights caused a decrease in sunshine on the street and onto the facades. The Charter of Athens, concerned with comfort and hygiene, developed the rule of prospect. Essentially, it aims to limit the negative consequences of taller buildings by limiting the height of the constructions in accordance to the width of the streets. In the first instance, accompanying the increases in population density, minor increases in building height were acceptable. In a second phase, the heart of the cities became the epicenter for densification and at the same time the main place of commercial and service activity; the limits acceded that it was no longer a problem to cast a shadow across the ground floor level. This increase in height, combined with the top floor being slightly set back in relation to the alignment, still allows for the construction of an additional floor. Finally, in the center of large cities, it is considered that the commercial function of the ground floor is complemented by offices on the second floor.

The technological evolution of transport and telecommunications, as well as globalization, even the transition from a traditional market economy to a continuous flow network economy, have modified not so much the visible morphology of the street, but its infrastructure, underlying (integration of distribution networks, cabling, optical fiber, connection possibilities, etc.) and its social use through the proliferation of socio-technical devices that this infrastructure enables (e.g. mobile telephony, guidance systems and tourist information, ambient intelligence, teleworking and even social networks). This means that the spaces of sociability move elsewhere: away from the street.

1.7.7. *The square*

The cityscape is dotted with a large number of public spaces and buildings that are distinguishable by their appearance, nature and function. From the theater to the hospital, including the square (or plaza), the school, etc., these spaces constitute places of exception and difference in the urban fabric. They play a role on two levels: on the one hand, at the level of the city, insofar as they constitute a part of the structure, and, on the other hand, more locally, where they maintain new relationships with the street, the block and the plots. The public spaces listed below are those that cannot be absent from the city since they constitute the skeleton of its social purpose. Within this framework, the square is a purposefully built element. Often, it took the location of a square in the original "checkerboard" plan, initially symbolizing the power that imposed it. The square has had different functions over the centuries. From the space of political debate in Greek society, to the commercial use of trade in the Middle Ages, the square has seen, according to the times, its functions reinforced or diminished. Its function has always been to provide more air, more light in the midst of a more or less dense fabric that has allowed for grouping and therefore exchange. Generally, this makes it possible to open up the perspective for several monuments. Sitte (1996) explains that in the Middle Ages and the Renaissance, town squares played a vital role in public life and there was a fundamental relationship between these squares and the public buildings that often occupied them. Panerai (1988) insists on the exceptional character of town squares in comparison with streets, which are more banal. As far as its structure is concerned, various analysts agree with the interpretation of the square as a closed space. Some theorists such as Sitte provide the general characteristics of the square. Main squares must have a width less than the height of the main building, while the dimensions of the main square must be related to the size of the city, with all other squares being smaller in size. In the regulations organizing the establishment of cities in the New World, such as that of Philip II in 1573, it is specified that the square must have a length equal to one and a half times its width in order to be able to organize festivals and military parades. A distinction must be made between different types of squares, beginning with the picturesque square, which is generally the result of a widening of the roadways and which offers, in a tight composition, wider sidewalks, sunshine and a different light. The squares can be located on either side of a road or at a crossroads. In this instance, these squares usually have an irregular and unobvious shape. In parallel to these squares, which are the result of a non-voluntary widening of the street, there are squares created on the site of a block that may yet be built. In this case, the shapes can be square, round, polygonal or oval. In this second type of square, there is no longer competition between the built-up area and the public space, since each is clearly defined. The center is naturally free of traffic and available for communal activities. The square, linked to the monument that is located there, therefore has the function of a square or a setting. The treatment of the facades is the object of a particular study since they are differentiated from those of the other streets.

In the industrial city, squares generally lose their original functions. The only persistent function is the commercial one, but even in this case, it is only occasional. The civic functions of community gathering have disappeared to make way for parking lots or sometimes a void when the latter is done away with.

In the contemporary city, citizens have been able to reclaim squares that have regained their original function, both symbolic and social, thanks to city center development projects which, by moving parking lots underground or to the outskirts of the city, favor pedestrian traffic.

From a methodological point of view, we refer the reader to the work of Bertrand and Listowski (1984) who are still very topical and propose the modalities of studying through the following four aspects: perception, morphology, experience and history. The importance of this analysis lies once again in its forward-looking nature for the organization of the system of public spaces in the city.

1.7.8. *Unique elements*

The different elements of the city call for a more comprehensive analysis. Many theorists have interpreted the function of public spaces. Rossi (1966) describes these places as “unique elements”, that is, nuclei of aggregation and poles of attraction. They are identified with certain built elements, such as facilities, monuments, vertical elements (towers, belfries, clock towers, etc.), as well as with elements of the natural landscape which, by their morphology and positioning, constitute as landmarks (a mountain, a tree, the sea, etc.). The French word for landmark is “repère” which is described in the Larousse-Bordas dictionary as:

> n. m., from the Latin reperire, to find. The reference points mark any object allowing us to orient ourselves in space, locate something, evaluate a distance (Larousse-Bordas 2000, author’s translation).

They punctuate the city, making its structure more legible: they facilitate, through their visibility, their recognition and spatial location, the construction of the mental map that the user forms in their mind as they wander through the public space. Today, the technical devices of geolocalization have brought the importance of these elements for the orientation of the individual and their role as distinctive elements at the heart of emotional security and kinesthetic decision-making back into discussion.

1.7.8.1. *Facilities*

Rossi (1966) analyzes the city as the voluntary establishment of public facilities (places of worship, town halls, schools, etc.) around which residential areas are organized. This theory produces a policy of one-off interventions by the powers that be, replacing a more global policy. Therefore, the specificity of each city stems from the particular tension between these unique elements and the residential areas. Rossi distinguishes architecture from these urban art pieces, in the sense that art is an element that exists for itself, while even the greatest architectural monuments are intimately linked to the city. These places are the guarantors of permanence, the physical signs of the past. They must be endowed with a certain quality, and Milizia describes them as having to be designed for the public good, well situated, and constructed according to the laws of suitability in the sense of the signifier and the expressive (Kaufmann and Bernier 1963).

Throughout the eras, churches have hosted the meetings of the city's inhabitants, long before the common house. They dominate the city with their size, height and architecture, which distinguish themselves from the modesty of other constructions. Their edifice is thus the object of a great financial sacrifice. Their establishment is never due to chance and its orientation, the altar, is carefully placed so as to face eastwards, that is, towards the rising sun, the dawning day, life.

The City Hall is sometimes housed in existing buildings; sometimes, it houses the belfry. If it is built on purpose, its architecture is very prestigious, but its plan remains simple. Town halls were usually built in a central location, often located on the market square. Sometimes, they contained a hall on the ground floor. This privileged position is symbolically necessary because of its role in the city.

Sometimes associated with the town hall, the tower, alternatively known as the *belfry*, had the function of alerting inhabitants to attacks, fires, etc. The belfry was an element that was often incorporated into the town hall, but was sometimes built alone, due to a lack of funds. Town halls thus had three primary functions: administrative, protective and temporal (work and curfew), as well as being a symbol of power.

Schools have not always been the object of a specific construction. They were first installed inside parishes, then town halls and finally as their own buildings.

The element that has long delimited the city is the wall. For several centuries, the enclosure was an integral part of the elements structuring the city. It has the double function of being defensive and symbolic. The role of this second function is to mark the differentiation of the city from the countryside and impose its power on its occupants. Its construction had a double effect on the physiognomy of the city. It led to the phenomenon of densification by preventing expansion when the population grew. It continues to mark the city even after its disappearance since it often gave way to ring roads.

Finally, the *cemetery* is today a big consumer of space; it has been gradually excluded from the city. Originally, the cemetery was attached to the church and this space participated in the city. This was no longer the case when its establishment became a consideration of the zoning concept.

Sitte (1966) makes a series of suggestions for the development of these spaces, considered as unique elements. He indicates, for example, the need to develop public gardens, at equal distances from each other, not open to the streets, but surrounded by houses or groups of public buildings. He insists on the effects of perspective, of the conservation of uneven terrain, of thoroughfares, of existing watercourses, and concludes on the indispensable irregularities. Unwin and Chabard (1922) advocate the need to "mark" the city centers by establishing a relationship between the different parts of the compositions. For these authors, a city of scattered buildings does not give any impression, since in ordinary streets they will not give us any overall architectural effect. The buildings put in relation to each other are mutually supportive, and the total result can indeed capture the imagination. The need to mark the city center should not prevent the development of secondary centers for cultural events, sports, etc., but the compositional remarks of the main centers retain their value for the latter.

1.7.8.2. *Monuments*

The autonomy of buildings and architecture in relation to the city began in the 19th century with the isolation of monuments, which, if still placed in an urban situation (squares, forecourts, perspectives), freed themselves from the plot constraints which the rest of the production remained subject to.

The future of the monument is linked to that of the urban fabric. Belmont (1987) warned of the error committed by confusing the country monument and the city monument. In the first case, an isolated building is built in the middle of nature, and "the city" then assembles around it. It is thus an "object monument". In the second case, it is inserted into the existing city, and the monument thus becomes an "urban space" into which different buildings are integrated. Most of the mistakes we have made stem from a confusion between these two approaches (Panerai 1988).

The monument should not impose itself artificially on the citizen, but should simply express a scale of value within the existing institutions, whereas it is the city that should be the object of a global monumentalization.

With the Modern Movement, we are witnessing a general monumentalization of construction, in particular social housing, which provides the bulk of the order. This monumentalization goes hand in hand with the segregation of activities and the isolation of architectural forms. The "correct and magnificent play of volumes under the light" (author's translation) seeks its justification in the naive hope of a new distribution of land ownership, where the land, entirely controlled by the community, would be accessible to all. In doing so, it ignores the concrete mechanisms of the production of the urban fabric and the appropriation of the built environment by its inhabitants (Panerai 1988).

The philosopher Debray (1999) defines several types of monuments from the type of reception of the public: the *message monument* that has communicative value, the *form monument* that has an aesthetic value and the *trace-monument* that has a documentary value.

Today, monuments acquire a special status: that of buildings or of urban and landscaped sites to be preserved, in a logic of heritage development, in order to pass them on to future generations. Debray's reflection therefore deserves to be pursued, in particular in view of the changes in perception brought about by advances in technical developments and digital technology. The mediation between the individual and the monument shifts towards media coverage that takes place via digital devices. If, on the one hand, these devices allow for a wider cognitive appropriation and an increased accessibility by different audiences, on the other hand, they have the consequence of distancing the user, by changing the temporalities of their reception which can be immediate or deferred.

1.8. Conclusion

The aim of this chapter has been to provide critical elements for understanding the city, through a morpho-sociological approach. In a logic of decomposition, we have therefore proposed some lines of thought, from which the elements that make up urbanity (both the morphology of the city, its meaning and its practices) can be analyzed. Their recomposition in a holistic system will bring out the meaning of the city. A better understanding of these elements is an operational tool for thinking, or even, rethinking the city of tomorrow by integrating the changes brought about by the different transitions we are experiencing: ecological, economic, digital, technological, societal, etc.

We have chosen to define the different elements according to a diachronic axis, trying to emphasize the key facts, the turning points that have impacted the urban morphology of cities, especially European cities, as we have inherited them and as we perceive them today.

The need for a paradigm shift in thinking on the development of our cities has emerged, as we have explained. At present, it is no longer a question of relying on economic attractiveness to think about land use planning; cities must be rethought in terms of the "harmonious living" of their inhabitants, through a circular and iterative approach based on the economy of contribution and networks.

1.9. References

Accorsi, F. (2010). *L'îlot ouvert : Christian de Portzamparc*. AAM Paris, Brussels.

Bachelard, G. (1966). *La philosophie du non*. Presses universitaires de France, Paris.

Belmont, J. and Architecte, F. (1987). *Modernes & [et] postmodernes*. Edition du Moniteur, Paris.

Benevolo, L. (1995). *Histoire de la ville*. Parenthèses, Marseille.

Berque, A. (1993). *Du geste à la cité. Formes urbaines et lien social au Japon*. Gallimard, Paris.

Bertrand, M.-J. and Listowski, H. (1984). *Les places dans la ville*. Dunod, Paris.

Braudel, F. (1979). *Civilisation matérielle, économie et capitalisme, XVe–XVIIIe siècle. Les structures du quotidien : le possible et l'impossible*. Colin, Paris.

De Vigan, J. and De Vigan, A. (2014). *Le grand Dicobat : dictionnaire du bâtiment 2015*. Arcature, Paris.

Debray, R. (1999). *La confusion des monuments*. Gallimard, Paris.

Dubois-Taine, G. and Chalas, Y. (1997). *La ville émergente*. L'Aube, Paris.

Fujita, M. (1989). *Urban Economic Theory, Land Use and City Size*. Cambridge University Press, Cambridge.

Genestier, P. (1988). Forme urbaine ! ... Formes urbaines ? *Villes en parallèle*, 12(1), 8–17.

Hall, E.T. (1966). *The Hidden Dimension*. Doubleday, New York.

Huet, B. (1986). L'architecture contre la ville. *Architecture Mouvement Continuité*, 14, 10–13.

Jaillet, M.C. (2004). L'espace périurbain : un univers pour les classes moyennes. *Esprit*, 40–62.

Joly, R. (1988). À la recherche de la forme urbaine. *Villes en parallèle*, 12(1), 18–41.

Kaufmann, E. and Bernier, O. (1963). *L'architecture au siècle des lumières : baroque et post-baroque en Angleterre, en Italie et en France*. Éditions Julliard, Paris.

Krier, R. (1975). *L'espace de la ville : théorie et pratique*. Archives d'architecture moderne, Brussels.

Laterrasse, J. (2016). Urbanisme et trame urbaine : ce que nous apprend l'histoire des villes. *e-Phaïstos – Revue d'histoire des techniques/Journal of the History of Technology*, V-1, 2016.

Lavedan, P. (1959). *Géographie des villes*, Volume 9. Gallimard, Paris.

Lavedan, P. and Hugueney, J. (1952). *Histoire de l'urbanisme*, Volume 1.2.3. Henri Laurens, Paris.

Lynch, K. (1969). *L'image de la cité*. Dunod, Paris.

Mangin, D. and Panerai, P. (1999). *Projet urbain*. Parenthèses, Marseille.

Mumford, L. (2011). *La cité à travers l'histoire*. Agone, Marseille.

Panerai, P. (1988). L'étude pratique des plans de ville. *Villes en parallèle*, 12(1), 100–109.

Panerai, P. and Mangin, D. (1988). *Les temps de la ville : l'économie raisonée des tracés urbains*. École d'Architecture de Versailles, Versailles.

Panerai, P., Depaule, J.C., Demorgon, M., Veyrenche, M. (1980). *Éléments d'analyse urbaine*. Archives d'Architecture Moderne, Brussels.

Panerai, P., Castex, J., Depaule, J.C. (1997). *Formes urbaines, de l'îlot à la barre*. Parenthèses, Marseille.

Paquot, T. (2020). *Mesure et démesure des villes*. CNRS Editions, Paris.

Pinson, D. (2014). *Actes du séminaire de design urbain, 14–15 avril*. IUAR, Aix-Marseille Université, Marseille.

Poëte, M. (1929). *Introduction à l'urbanisme : l'évolution des villes, la leçon de l'antiquité*. Boivin, Paris.

Riboulet, P. (1998). *Onze leçons sur la composition urbaine*. Presses de l'École nationale des ponts et chaussées, Paris.

Rossi, A. (1966). *L'architettura della città*. Marsilio Editori, Padua.

Sitte, C. (1996). *L'art de bâtir les villes. L'urbanisme selon ses fondements artistiques*. Point-Seuil, Paris.

Unwin, R. and Chabard, P. (1922). *L'étude pratique des plans de ville*s. Impr. Protat Frères, Macon.

2

Space, Production and Urban Forms

Jean-Pierre MIGNOT
LERASS, UPS, Toulouse, France

2.1. General introduction

The concept of space occupies an important place in the general history of thought, in the same way as time, so much so that it is presented differently according to the disciplinary fields which employ it and imbue it with meaning. Nevertheless, here we use a generic definition of space as an expanse in which certain objects are positioned. Space is thus a medium which is characterized by the simultaneous and external existence of the parts which constitute it, and which therefore allows us to grasp these same objects, localized and perceived, and this observation of bodies (objects) is well and truly situated in this expanse. Therefore, the city as a social body certainly appears as an artifact and also as a perceived body (object), expanding within a space which contains it.

Two qualifiers emerge from this first overall definition of space and the city it contains: on the one hand, the city as an artifact, and on the other hand, space as a place where this unique production that is the city takes place. However, the perspective we have on the city and on this expanse that contains it is itself unique, insofar as distinct disciplinary fields each elucidate diverse meaning. If the space in itself tells us nothing about itself, the different ways we perceive it and the objects (bodies) that compose it will explain a lot more. There is nothing natural about the city, and it is above all the product of the techniques used in its manufacture, employed in its construction, and as such, the relationship with the space that contains it is necessarily interesting for understanding one in terms of the other. Indeed, if the city as a technical construct (even if it cannot be reduced simply to that) is the result of human production at the same time as it is a production of space, because once characterized by the object it contains, the space in turn becomes a production; the space for production on the one hand, as well as the production of space on the other. As such, these are the elements that should be considered in order to highlight the main characteristics of one and the other.

At this point, it is necessary to understand the interdependence between space and production in general, that is, the way in which these two structurally linked elements overdetermine each other. For example, the city of the Middle Ages and the modern city do not have the same functionalities, of course, nor the same structures, themselves dependent on the conditions of production that characterize the system to which they belong and within which they have developed. Beyond a common definition to characterize them, the forms they take will essentially depend on the production conditions in which they are produced. This differentiation, which is the product of particular historical circumstances, therefore refers to the way in which the city is produced and the ways in which the economic growth that presides over its development has itself been instituted: the forms and functions of economic development in particular, and societal development in general, will therefore greatly interfere in its construction. In the analysis of the city and its development, it is therefore appropriate to associate the reasons why this development occurred in a certain way and not in another: we propose to link the modalities of this growth, that is, the different forms of accumulation, to the forms that the cities took during this development. In other words, the city and its development are strangely similar to the development of the economy in which it is embedded.

This is how, in our opinion, it is possible to account for the social impacts of urban development on the environment in which the city is located. By way of example, as we will see, "the ring city" can only emerge out of an economic system in which the car is king, and in order to do this, this consumerism must be en masse. Added to the phenomenon of the car is its demographic pressure, particularly significant in terms of its economic development, especially with regard to its "sustainable" aspect: in fact, by 2050, it is projected that 80% of the world's population will be living in cities. It is not hard to imagine the challenges facing such a transformation.

As a result, it is easy to understand that the environmental impact of such development is different depending on whether the means of transport uses a carbon-based resource or animal traction. What then becomes of the city in a world saturated with pollutants, and moreover, how to orient the city of the future towards an existence that is respectful of this environment and its populations? Can the city, which itself contributes to the destruction of its environment, suddenly become a regenerating factor for this same environment? How can we ensure that the growing number of megacities with more than 20 million inhabitants, which are not equipped to provide livelihoods that are respectful of this environment, will be able to support a population that is itself a victim of these abuses? Closer to home, the very recent importance of "regional metropolises" draining an already distressed rural realm, poses the problem of balancing the local, which is on the way to extinction, with the global, which is palpably too invasive, particularly through its form of peri-urbanization. This is what we propose to analyze at the end of this chapter, by questioning the resilience of the city in the face of these challenges.

2.2. The city between space and time

We start from the founding hypothesis that the city appears to be an important and now dominant mode of occupying space, with particular regard to the rapid economic and demographic developments that the world has experienced[1]. That said, such a hypothesis is not sufficient, at least not until we have resolved the enigma of space, which we formulate as follows: what is space? Or, more precisely, is there an epistemological status of space as a concept? The question of the city which is etched into this space thus takes on a distinct relief with many ontological contours, which will, at the same time as the contents, the meaning, effects and consequences, need to be put into perspective: the city, initially perceived as an expressive and

1. Here we refer to the developments of Western economies, of course, but not exclusively. In fact, in recent decades, emerging countries are the perfect example for these urban developments since we can see that the largest megacities are in fact situated in these countries. For example, Sao Paulo, Shanghai, New Delhi, Lagos, etc. Moreover, in this respect, these megacities have developed more rapidly than those in countries such as France, the United Kingdom, Germany or Italy. None of these countries experienced such rapid urban developments.

technological manifestation, therefore takes on its entire meaning, and with it the technology itself.

2.2.1. *Back to the concept of space*

An initial observation must first be made: if the "hard sciences" have expounded a definition of space that is both clear and precise, the same cannot be said of the humanities and social sciences (HSS). Indeed, when we want to locate a point in space or the position of an object within a set of coordinates, it is necessary to locate it via other points or other objects, thereby becoming a reference system: a Cartesian coordinate system expresses this idea perfectly. Subsequently, physics itself will question the ideas freely inherited from Aristotle and the Aristotelianism that is based on the belief in the existence of an absolute space. Now, 450 years after the Copernican revolution and the affirmation of the relativistic ideas of Galileo, Aristotelian thought remains well rooted in the great minds of the time, in its belief of an absolute space, as well as of an absolute time. With Einstein's theory of special relativity, space loses its status as an absolute concept: the positioning of two observers, themselves in unison relative to motion, implies that there can be no absolute space. It therefore seems appropriate that, in order to better understand the relationship between space and the city, we must first return to the concept of space. Since the concept of space contains and determines in a specific way the forms that the city takes, in both time and in space, obviously, both these fundamental aspects that appreciate what the city is, will be examined at the same level. To begin, the concept of space will be addressed first.

2.2.1.1. *Space as a concept: from physics to philosophy*

The different conceptions of space appear in a double form: on the one hand, a so-called "absolute" space, which exists in and of itself, independent of the forms and bodies (or objects) that it contains; on the other hand, a "relative" space, which is only understood by its relations to the forms and bodies (or objects) that it contains or maintains within itself. First, the notion of space was, for all intents and purposes, a matter of physics, that is, natural philosophy, and therefore only yielded a physicalist approach[2]. This is how

2. From the Greeks onwards, the concept of space only takes shape within the framework of physics. Moreover, one often compares the place at which the parallel concept of time was considered, the place of space becoming, from the beginning, comparatively less important. See Aristotle (1990). Concerning space and its conception, also see Plato (2016).

Einstein opposed the two forms of space, one that he qualified as a "local or topological quality of the world of material objects"[3], the other being conceived as the "receptacle of material objects". In this way, by distinguishing the bodies from the support in which they move, we logically arrive at the existence of a unique support, space.

In any case, space is thus understood as the instance within which the different movements of objects (or bodies), the things that compose it, occur. The issue of a definition of space therefore appears first of all as "kinetic", that is, as the topological place where movement occurs: it is therefore a question of explaining, before anything else, the way in which the objects of which it is the receptacle move, of explaining this movement, and making the definition of a known system of reference necessary. In this way, the general concept of space can (and must) only be emphasized within the framework of the social sciences, only if the spatial character of objects is itself defined and conceived. In this respect, we cannot conceive of this space as a simple expression, a manifestation of a Euclidean space[4] within which objects would be in motion. Moreover, the notion of space must define the morphology of the objects it contains as the basis of its understanding. In fact, it is from this perspective that we approach the city as a privileged object in terms of its occupation of space.

Philosophy, starting with Descartes, contributes to the development of answers to the questions that the concept of space poses. Indeed, from the analysis of space that Descartes proposes, which is developed under the term of "extended", it begins from the principle that "corporeal substance cannot be clearly conceived without its extension"[5] This corporeal substance, namely matter, is homogeneous, as Galileo explains, but its profound nature lies in the fact "that it is extended"[5], that is, that it is "space". Therefore, there is no vacuum in the world (unlike for Newton) since we know the motion of a body by measuring it locally, that is, by appreciating the way it moves from one stationary position to another. This implies that Descartes

3. See Jammer (1993), author's translation.

4. Contrary to Newtonian physics: insofar as absolute space, as defined by Newton, is based on the complete independence of space (and time) as separate from matter. It is an absolute space, empty of all bodies and which remains so, even in their absence, thereby making Newtonian space a three-dimensional Euclidean space, an infinite and unchanging space. See, in particular, Newton (2016).

5. We refer here to the title of paragraph 9 in the second part of "Principles of Philosophy". For more on this subject, see Descartes (1953).

identifies matter with the expanse[6] that contains it, notably by establishing the existence of an indefinite space, itself elaborated in the form of a Euclidean space (length, width and depth) and separated from the bodies themselves. This separation occurs only through the act of thought because there is no real difference between the material substance and the extension (space) itself.

If Descartes extends the discourse of physicists in a certain way, notably in the debate with Newton and Leibniz, as far as Kant is concerned, the latter develops an approach to space (and time) which more specifically questions what knowledge is and the conditions necessary for its emergence. Space (and time) constitute "pure intuitions"[7] and a priori data of sensitivity, that is, intuitions that are, prior to any experience, free from the senses and independent of them (Kant 2021). Kant looks for "the conditions of the possibility of experience", whether these are internal and/or external, namely what makes any experience possible for the human mind. He takes space (and time) as a priori forms of knowledge, because the principle of what emerges from space in the representation already exists in the mind prior to any experience. Therefore, space, which would be neither an empirical data nor an intellectual creation, becomes a knowledge that participates in the construction, a priori necessary for the synthesis between experience and understanding. Under these conditions, space (and time) becomes the condition of the knowledge that associates experience and understanding.

Finally, we will discuss a third conception of space that seems to us to be the appropriate one for understanding the city as a major instance of spatial occupation, which we will discuss in greater detail later on. Leibniz, pitted against Newton, submits the definition of space to that of the order that exists between the bodies that constitute it, as well as to the relationships that these same bodies maintain between them (Robinet 1957). Going against Descartes, he criticizes the separation of matter (and its corporeal extension) and spirit (or soul). Space appears as an order of things, a general ordering link for the objects it contains. Therefore, "nature" cannot be considered as a pure geometrical space, and if we want to hear and understand what an object is, what a body is, we must make the hypothesis that a positioning in

6. See Descartes (1953).

7. For Kant, intuition remains a mode of immediate knowledge, that is, without any possible intermediation, but moreover, it owes nothing to experience. Consequently, it is not the intuition of anything, even if it determines a priori a form of sensibility, an intuition of all things.

space is the consequence of the constitution of its nature. Therefore, if matter divides infinitely, the body cannot find any unity in its parts. This unity therefore depends, says Leibniz, in his principle, solely on the primitive force that animates it (Leibniz 1994).

If we wanted to develop (at least) the different conceptions of space, it is because space is filled by the city and this relationship between the city and space is fundamental to the proper positioning and understanding of urban problems. Admittedly, the city is an artifact, but it is charged with history, meaning and therefore with a problematization that cannot be ignored if we are to appreciate that it cannot be reduced to this approach alone. We will see later that the position one gives to the city (and to the urban world in general), is close to this vision of space that Leibniz extols, insofar as the legibility of the city as a unit also depends on this layering, the result of its history, its different parts and constituent elements, that allows us to read this space/city relationship like an open book.

2.2.1.2. *Space and the humanities and social sciences (HSS)*

The explanation of space cannot be an end in itself because, like time, developing such a concept requires setting it into motion, thereby demonstrating its relative importance. The city as the main organ of space occupation[8], as a "body" that is positioned in space in a privileged manner, as an order (in the sense of Leibniz) whose parts, before forming a more or less homogeneous whole, unfold according to a mode of representation, taking us from the representation of space in the urban form, to a space whose representation is a reflection of society and its mode(s) of representation.

2.2.1.2.1. The representation of space and the space of representation

The representation of space

In the approach of space that we propose here, we cannot fail to question the ways by which we represent space, that is, the multiple ways in which the object of thought, here space, marks its presence in our mind, depending on the way in which it is presented. Space appears then as this instance which contains these objects in the manner Leibniz described it: the city thus constitutes this object which contains space and its mode of representation is

8. Recall that by 2050 more than 80% of the world's population will be living in cities.

multiple. We will choose to illustrate it through the cartography of our space which becomes emblematic of the mode of representing space. Indeed, from Ptolemy[9] to modern satellite cartography, we begin to discover, through this form, an increasingly precise reality of our living spaces. With the cartography of the earth, we observe an ordering of space, its development, with particular regard to the objects that constitute it. This is especially true of the urban system. Thanks to the plotting of latitudes and longitudes, or triangulation[10], territories have been slowly exploited over the course of this slow evolution in the representations of space which has led to a universal system whose modifications and evolutions can today be followed in real time. This representation of space is all the more important since it allows us to follow changes that are linked to urban development, from a physical point of view, a demographic and economic point of view, etc.

The space of representation

An important aspect to understand in the discussion on the space of representation is that we can draw the map of the world, and therefore the cartography of humanity as a whole, on a single sheet of paper. This configuration, both exhaustive and unreal, takes the form of the reality and the virtual. The world that is drawn before our eyes is indeed the one that really exists, but at the same time, it is only its representation, namely, the space of its representation, as real as it is figurative. If this is the case it is because we succeeded in transferring the physical aspect of the world to a representation of its space, reduced to its simplest expression through the implementation of a codified language which delimits and shrinks this space. We therefore recall here the definition that Marion Ségaud proposes:

> from a set of socially accepted means of codifying, at a given moment, the way of representing space; it is based on conventions resulting from the technical and social division of

---

9. These representations are developed on the model of Ptolemy, who developed a system of projections based on the geographical calculation of latitude and longitude coordinates. However, it is not until the great discoveries of the 16th century that it becomes possible to locate a point in space that represents a city, a coast, a river, etc.

10. Triangulation is a term used in geometry and trigonometry. It is the technique of determining the position of a point from the angles between that point and the other reference points whose positions are known. It can be, for example, the point of a vertex of a triangle for which we know the other two angles and the length of one side. This technique is also used by astrophysicists in GPS guidance technology for accurate positioning in three dimensions.

> labor. The space of representation is an instrument born of social constraint (Ségaud 2009, author's translation).

It therefore refers us to a projection of reality that indicates, to us, the social forms of representation and the technical conditions of these same representations, which are the fruit of the technologies implemented during the reference period. In this way, in the end, what separates the representation of space from the space of representation is the fact that we can differentiate the setting in the form of reality in the first case, from its virtual appreciation in the second.

2.2.1.2.2. The limits of HSS in grasping the concept of space

The various analyses and reflections concerning space and its concept in the field of the Human and Social Sciences really developed during the 1960s and 1970s. The "geographers" in many respects will become pioneers in this field, even if they make space a secondary concept whose objective is to serve as a base of reference for other processes that are deemed to be important, such as the phenomena of sociological, economic and historical nature. This means that these phenomena take shape in space and that space becomes the determining place within which form takes shape and becomes a tangible reality. In this way, the definition of space remains a place where a social reality shows itself, and that a space, even if it is simply the receptacle of a reality other than itself, remains… a space.

That said, geographers were the first to think of space in two ways: on the one hand, in terms of a "physical" approach that gives it a fixed, natural quality. In this way, the continents remain continents, the division of coasts and tectonic plates are always invariants, with the exception of modification by nature itself (erosion, for example). However, at the same time, space is considered in terms of a "power relationship" between the different elements that constitute it, and if space is first a physical entity, it also remains a cultural reality within which civilizations, for example, are posed and opposed. On the other hand, the geographer distinguishes different cultures, different socio-cultural spaces as absolute realities: for example, Western civilization, South American peoples, etc., even if, as a system, the realities that characterize them remain and the elements that constitute them in their reciprocal relationships are never predetermined.

In any case, what characterizes the geographer's approach (Di Meo 2005) makes sense first and foremost in terms of the consideration given to space insofar as this remains their privileged field of intervention. As such, it is easy to understand that this academic field has become a priority for understanding space.

The economy approach: a model of spatial dynamics and its limitations

If we stick with that which the neo-classical (i.e. neoliberal) approach proposes with regard to the consideration of space, regardless of its form, we would have to admit that it enters into the construction of the discourse like an intruder. Indeed, the neoliberal analysis, described as orthodox, sets up a system that determines the conditions for fixing an optimum, so long as the "natural" conditions for the functioning of markets are respected. The economy is thus in a situation that is not only balanced but stable (Arrow and Debreu 1954). This equilibrium is the result of adjustments that the free functioning of the market allows within the framework of the price/quantity relationship, so that any change in the quantities exchanged would lead to an adjustment of the prices, in order to maintain an overall equilibrium between supply and demand. These adjustments are made instantaneously and at no cost to the agents. The result is that, on the one hand, the demonstration is complete and self-sufficient in terms of determining an optimum and an automatic return to a stable equilibrium, and on the other hand, the problems of equilibrium are avoided in cases where the optimum cannot be demonstrated, because in these cases, it is the model itself that is radically in question. As we can see, there is no place for space as we have defined it above, which we can delineate in different forms, such as territories, regions and nations.

Therefore, other economic approaches have developed over time based on the consideration of space and the differential factors these elements could bring about (Paelinck and Sallez 1983). In this context, we will only mention a few examples to help highlight this perspective.

Macroeconomic development models and the introduction of space

The growth years of the so-called "Fordist" era saw the emergence of a theoretical debate in the form of the macroeconomic modeling for the conditions of growth and its durability over time[11], and which are based solely on developed economies and their economic evolution. This means

11. We refer here to the various growth models developed during the 1960s and 1970s by authors such as Domar, Harrod, Kaldor, Solow, Samuelson and many others. For further information, we refer the reader to the exceptional reference work of this field by Hahn and Matthews (1972).

that the positioning of the authors' problematic is immediately identified with a specific space, that of developed countries, and more specifically that of Western economies. The negative aspect of this was the appearance of problems linked to underdevelopment, that is, the existence of a differentiated and qualified economic space as a specific space. As a result, the dialectical dualism of wealth/poverty has been associated with the dialectical dualism of development/underdevelopment as well as, and more explicitly, with the spatial and geographical relationship of center/periphery. Because the poverty of the so-called "underdeveloped" countries is only relative to the wealth of the so-called developed countries, the difference is also, perhaps even more so, expressed in terms of space[12].

Many voices were raised against the theoretical approach coming from Western circles, which emphasized the conditions of growth and, above all, showing the way forward for economies whose history had been excluded from this historical phase. Criticisms were leveled against this vision, which was considered to be too linear[13]. Critical analyses were to develop on the same theme by authors influenced by Marxist critiques, most of them putting into perspective the contradictions, the conditions and the consequences, often negative, of economic growth in the space of contemporary capitalism. In this respect, and in line with such a logic, Alain Lipietz puts into perspective the destiny of capitalism to establish a new "inter-regional" order, notably by redistributing production activities within its own territory. In other words, space in the form of territories undergoes the logic imposed on it by capitalism, in particular because of the logic it puts in place (disintegration of the rural economy, deskilling of work, decline of traditional local industries, etc.). We note the premonitory character of certain analyses, particularly in the context of the claims of the French "*gilets jaunes*" (yellow vests) movement, with regard to the abandonment of territories, as well as, and we should emphasize this, these different authors put into perspective the ability of existing productive systems to destroy their environment.

In fact, what we want to put into perspective here, admittedly in too limited a framework, is the importance, the richness and the scope of taking into account space in the economic analysis, as it facilitates the understanding

12. See also Aydalot (1980).

13. The main critics on the economy of the period, articulating arguments around the "unequal exchange" of a "world economy" were: Emmanuel (1969), Baran (1967), Baran and Sweezy (1968) and Lipietz (1977).

of global mechanisms due to its close proximity and concern for reality[14]. Rather than sticking to a superimposition of spatial models on the one hand, and economic models on the other, it would be more appropriate to give meaning to a synthesis capable of putting into perspective the conditions for the adaptation of an economic system to the introduction of unique spatial constraints (Aydalot 1985).

Space and the productive system

It seems that it is through the specific approach to productive systems that the consideration of space has been the most complete and best simulated. As the author indicates,

> the main problem facing these new approaches probably lies in the necessary but often difficult transition between the concepts of industrial fabric or space and the representation of the productive system that must now emerge.[15]

We find here the problem of representing industrial space within the synthesis framework between the "terrain" (or the place) and the productive

14. A necessary note here: the neo-classical model that we mentioned above perfectly illustrates the fact that it substitutes an explanation for the functioning of reality, which is the basis of scientific analysis, a purely normative problematic that is of little interest outside the logic of "defense and illustration" for the functioning of the market economy. In this case, the use of mathematics cannot constitute a guarantee of scientificity. We will illustrate this point with the following example: at a given point in the labor market, any change in the supply of labor (on the part of employees) and/or in the demand for labor (on the part of employers) will result in a change in the price of labor, that is, in wages, so that the suppliers will react to a change in the equilibrium variable. If supply exceeds demand, wages will fall because the ratio of capital to labor decreases, which implies a fall in the marginal productivity of labor and thus in wages. In this case, there is a fall in supply on the part of employees, which leads to an increase in salaries due to increases in the capital/labor ratio, and with it, the marginal productivity of labor and therefore wages. This mechanism should continue until an equilibrium point is determined. At this point, if we introduce space into the mix, we can then hypothesize that in two neighboring spaces (e.g. two neighboring regions), there are different levels of marginal productivity, that is, different wages. In this case, the adjustment will not be made through salaries but through migration from one region to another. We now know the importance of these movements between underdeveloped and developed countries, which perfectly illustrates these situations, because at the same time they discredit the liberal theoretical construction, which proves incapable here and elsewhere of introducing the concept of space. We could also add more examples.

15. See Torre (1988).

system. A description of economic relations that takes space into account cannot forgo an industrial analysis of the productive structures (or systems) involved, since the local, regional, national and international environments of these different productive systems cannot do without this fundamental determinant that is space. However, two seemingly contradictory observations can be made: on the one hand, it can be seen that certain investments can modify local (or regional) industrial structures and, as such, they play an important role in the development of local (or regional) productive systems; on the other hand, as certain authors[16] strongly emphasize, the deficit in public policies and the abandonment[17] by the state of territories furthest away from productive centers (and therefore from employment areas), has the effect of provoking a fragmentation not only of local productive systems, but, above all, impacts the social fabric, and therefore spaces are affected. For many decades, the practices of the state have made it possible, thanks to the transfers made between these different spaces, to maintain a certain equality between rich, privileged areas and poor areas. However, with the effects of globalization on local (or regional) productive systems, on the one hand, and the problems associated with debt resulting from disorderly deregulation on the other, things have changed considerably. Therefore, the crucial question for the future is: should we resolve to choose between (relative) equality between territories or economic efficiency? This question alone highlights the importance that space will play in the future. If we add to this already somewhat tarnished picture the condition constraints imposed by an unavoidable ecological transition in the long term, we can logically and without difficulty envisage the importance of the problems to be solved, all in all in the fairly near future. Of course, many solutions have already been developed and are being debated[18].

As we can see from this albeit limited overview, the concept of space is important for putting into perspective the construction of discourses (in the Greek sense of "*logos*") and therefore in the cementing of knowledge and the determination of knowledge. Whether it is a question of economic thought or the construction of geographical models (we are thinking here of the establishment of networks and the relationships that are established that make the so-called "world" economies work), space is in all cases unavoidable

16. We refer in particular to the works carried out by Laurent Davezies (2012, 2015).

17. In this respect, it can be noted that these challenges are largely in line with the demands made by the "yellow vests" movement, which has from the outset called for a radical overhaul of inter-regional solidarity aimed at a better distribution of wealth within the national space.

18. Among other authors, we refer to Veltz (1994, 2017).

and certainly cannot be dismissed without considerably weakening the scope of the analyses which factor it as an indicator.

2.2.1.3. *From space to the city*

We have just seen that the concept of space occupies an important place in the history of thought, and also as an essential concept in the construction of knowledge, especially since it is declined in singular modes according to the various academic fields that put it in motion and give it meaning. Because it is an "expanse" which "contains" certain "objects" or "bodies", space appears as a medium that is characterized by the simultaneous existence of interior and exterior parts (or elements, or bodies, etc.) that constitute it. In this respect, it allows us to locate the observed and perceived objects, which amounts to the glance/gaze of these bodies or objects within the framework of this "expanse". Therefore, the city appears as this "object", this "body" which extends within the finite space that contains it. Two qualifiers emerge from this initial global definition of space and the city: the city as an artifact and space as the place where this production is accommodated (for every artifact is the result of a production of man by man). However, the way we look at the city (independent a priori of this "expanse" which contains it) is itself unique insofar as it belongs to distinct disciplinary fields (in the sense of Bourdieu), each of which sheds light on its meaning and scope. Because space in and of itself does not indicate anything of itself, the distinct and differentiated ways in which we look at it and the objects (or bodies) that constitute it tell us much more. We will therefore know this space all the better if we understand the objects it contains, especially when it comes to the city. Hence the idea of having to consider space not as a fact "of nature", nor as a fact of culture, but first and foremost as a produced whole.

As we have said, the city is first and foremost a technical construction and is, as such, an artifact. To understand the relationship between space and the city therefore requires an understanding of both. Indeed, if the city as a technical construction (even if it is false and vain to want to reduce it to a simple artifact) is a product of man, it is a production in space and, as such, space is a space of production, that is, the production of the city. In this respect, the centralizing state plays an essential role in setting up the conditions for an organization dedicated to space, that is, the city. Therefore, under the aegis of the state, various organizations (and therefore agencies) and institutions whose objectives are strictly limited to the city, to urbanity and to its development, begin to emerge. Whether it is the delegation for "land use planning" during the 1970s and 1980s, or the various "city" ministries created to respond to suburban problems, the objective of producing a regulated and socially acceptable urbanity has been a constant for decades. Nevertheless, because space is also identifiable with the objects it contains, we must admit that these different contents are themselves a production of space. The space of production on the one hand, and production of space on the other, are determining factors that we propose to examine closely in order to highlight the main characteristics of space and the city, both for their own sake and in terms of their reciprocal relations.

At this point, it is necessary to understand the interdependent relationships that are established between space and the production of the objects it contains, that is, the way in which these two elements overdetermine each other. Since the cities of the Middle Ages and contemporary cities do not have the same characteristics (apart from their common definition) because the forms that the productive systems that generated them are different. In all cases, the city as a product, and above all as a social product, cannot be considered as an ordinary product, as a simple object of consumption. It is in fact above all a totality, that is, a unity, a synthesis and, at the same time, the plurality of all the elements that constitute the whole. These same forms of cities, which have evolved in time and space, themselves depend on the conditions under which they were produced. This is the point that we now propose to address.

2.2.2. *Produced urban space and social production*

The city as an expression (as the concrete and technical manifestation) of a mode of production of space, and thus as an expression of a mode of accumulation specific to this mode of production, is a historical product and is therefore integrated into a unique historical perspective: there is no reason to say that a given period in history (in this field as in others) is more "intelligent" than another, except to ideologically connote the criteria of "intelligence".

Let us start with two particular hypotheses developed by Lefèbvre (1986) and Riboulet (1998): on the one hand, we think that the production of space leads to space as a product, as a commodity within the framework of capitalist production; on the other hand, the city as an artifact expresses the technical characteristics of the dominant mode of production, the development by productive forces, within which it emerges. We will develop these problems in two ways.

2.2.2.1. *From the production of space to space as a product*

We start from the assumption that the social space represented by the city is first and foremost a produced space. This seems obvious, but we need to take a closer look at the consequences of such a statement. In this perspective, we consider that physical space, the space that "nature" gives us, is not confused with what we call social space. The latter is not reduced to a simple display of social or even societal phenomena or an empty set that is convenient to fill. On the contrary, it constitutes the very foundation in which this human social whole that constitutes the city is inscribed. And if this specific space is made up of diversity, of multiplicity and of differentiated sub-spaces and divisions that may appear heterogeneous, it is nonetheless the result of a history, a history of humankind immersed in a lived social world and therefore in social relationships. Several remarks:

– This space is made of both opacity and transparency: it expresses at the same time the refusal of the naivety and that of the ideological disguise. It is as much the conditions of existence of the space and the way of organizing it that reveal the illusion of its substantial character and the weight that we grant to its "naturalness", while at the same time maintaining a certain mythology, even a certain form of mystification. Conversely, the supposedly acquired transparency very quickly becomes a kind of philosophical ideality that is ultimately devoid of meaning. In the end, grasping the essence of what space is amounts to trying to determine not only its constituents but also all of its concrete determinants.

– We voluntarily put the physical space at a distance even if it remains the "original sin" of the social space, that is, the socialized and therefore "societalized" space. Under these conditions, the physical space becomes a decoration and each object which belongs to this physical nature becomes a symbol to be protected. Naturalist approaches to sustainable development are a significant example: rediscovering the authentic, questioning a "raw material" upon which "sciences and technologies" have acted to produce this space in the course of destruction. Of course, this difficult relationship refers to the unfortunately granted wishes of Descartes who called on man to become "master and possessor" of nature.

Each society, each mode of production produces a unique, specific space, the characteristics of which depend on the technical qualities it contains, but which remain a dominant universal throughout its existence. For example, the ancient city is characterized by a specific spatial practice that shapes a space that is specific to it and that it has appropriated. The same goes for the cities of the Middle Ages. Indeed, each mode of production associated with its own production relations can be interpreted from studying the city, even by the significant variations[19]. The notion of social space does indeed imply a certain complexity that resists an overly linear, even mechanical, analysis. With the development of capitalism, the city takes on contours that the historical forms of capital accumulation imply. We will revisit this later on.

Therefore, social space as a product of history appears in all its complexity both from a formal point of view (e.g. mode of appearance of the different urban forms) and from a real-abstract point of view (articulation and meaning of these different forms). In this way, the urban space contains multiple interconnections, assigned to places and locations, which overlap and intertwine. In the same way, the representations of the social relationships of production, which determine certain superstructures of society, envelop the particular and effective forms of power in the space that are consequently inscribed into the buildings, the monuments, works of art, etc. across the cityscape. We find ourselves therefore within the framework of a triptych, which we can summarize as follows:

– on the one hand, a social practice that refers to the production (as well as to the reproduction and to the conditions of reproduction) of the relationships of production, that is, spatial sets specific to each social formation, which aims to ensure a certain continuity in time and space;

– on the other hand, at the same time, this presupposes a representation of the space that is consecutive to the dominant order and to the logic imposed by social relations, as well as the forms of domination which are attached to it, and also the evolution of knowledge and technologies, signs, social codes that emerge, including those that are confrontational.

Finally, regardless of the codes, the forms of domination, etc., we are witnessing the emergence, on the stage of history, spaces of representation within which "Art" appears as one of the possible codes of these spaces of representation. In this way, the representation of space and the space of representations mutually generate, articulate and reticulate within a complex system made up of more or less well-ordered superimpositions.

---

19. This assertion is perfectly illustrated within the framework of a morphological approach to the city, developed in Chapter 1 of this book.

2.2.2.2. *From production to product: semantic details.*

We have so far used the term production without formulating the precise meaning that we assign to it. The concept of production has two particular meanings, one broad and more "societal" in nature, the other restricted but socially determined.

On the one hand, taken in its broadest sense, production expresses the activity of humans as social beings who produce their life, history, consciousness, that is, their lived social world: the material and spiritual conditions of their life. Nothing appears in history and in society that is not acquired and produced. Nature is included and transformed by the simple fact of the continuous and powerful intervention of humans and is thus modified, and even subjected to the technical ordering produced by humans. Production in the broad sense therefore includes many diverse forms (ideological, philosophical, religious, political, etc.) as well as multiple works that may (or may not) bear the mark of producers. The city as an artifact, and in its general sense, as the fruit of these works that are produced on a reduced and specific scale that nevertheless represent an era, imbuing it with a unique meaning. It suffices to think of cities such as Venice and Florence, which remain the result of a production which are unique, not reproducible ad infinitum, as are commonly manufactured commodities.

On the other hand, in a narrower sense, production refers to the products themselves. It therefore takes on the common meaning that we give it, for example, through the economic language to which it is most often attached: Who produces it? How is it produced? What is produced? and so on. The gesture of producing thus becomes a universal gesture and this universality is found in the object which signifies, in the framework of the capitalist mode of production, the commodity. In this case, it is no longer a question of the production of a work of art, but on the contrary, of the current and massive production of goods destined for consumption and therefore indefinitely reproducible, as far as conditions allow.

In this way, humans, in their social practice, simultaneously create works that most often remain "unique", but they also produce things (whether material or immaterial) that are commodities and that the medial or direct consumption thereof contributes to destruction[20]. In both cases, it is the work of humans that is at the center of the act of production, whether this work is expressed on a social level or on an individual level. The concept of production thus remains a tangible universal by which, and thanks to which, humans impose themselves "as master and dominator of nature".

Therefore, to return to the question of space, it is not the localized space (in any case not exclusively) to which we are attached but to the produced space taken in its totality, in its globality, and which we must therefore approach from an analytical point of view at the same time as from a point of view that must entail and must be developed within the framework of a theoretical knowledge. At this point, we agree with Riboulet's approach (1998) vis-à-vis the urban composition.

2.2.3. *Mode of production, production relationships and urban composition: the weight of history*

As discussed earlier, there is no intrinsic nature of space and there is always an "already-there" from which any new division of space takes place in a spatial matrix that characterizes each mode of production, which, on the one hand, fit together in a more or less visible way, and, on the other hand, specify particular relations of production that characterize each of the modes of production concerned. As a manifestation of space as product, the city is also presented in an imperfect way as the result of this space-product or space-commodity. Consequently, the city as an artifact does not escape its destiny as a technical object that is a product of its history.

2.2.3.1. *The urban composition*

Defining urban composition comes down to talking about composition both in the aesthetic sense of the term and in terms of the city, that is, to refer to two elements which we readily associate but whose theoretical statuses differ greatly (without being contradictory). Indeed, while composition is the act of an individual, the "composer" who affirms their composition through their work, elaborated "on the blank page" after a slow and thoughtful work process that results in a finished object in time and space, the city is a collective product in which a multitude of individuals, communities and

20. Under these circumstances, an economist like A. Schumpeter spoke of "destructive creation". However, you may agree that with regard to these evolutions (or involutions), there exists a certain irony, albeit unintentional on the part of the author.

institutions participate, none of which (and all of which) can be identified because they are drowned in a mass of elements as diverse as they are multiple as well as complementary. Moreover, this multi-faceted city is never finished, always fluctuating, so much so that it seems difficult to analyze or even understand an object that is gradually slipping away and constantly changing.

Therefore, it seems that oppositions, ambivalences and ambiguities are the common trait between composition and city. Characterizing urban composition is thus tantamount to explaining a double movement of oscillation, which refers us in turn to aesthetic, formal visions and to socio-politico-economic approaches, like an open work into which we can enter even when it has constantly been repeated and restarted. However, approaching the urban by its socio-politico-economic aspect also consists of identifying the actors who take part and participate in its construction, in its evolution. We are then faced with social, economic and political forces which act directly and/or indirectly on its structuring in time and space, and which cannot be considered as blind forces acting in a disconnected and chaotic way. Clearly, they intervene in the process of composition (aesthetics) and give a unique form to the space while being part of the process that conditions the social reality of the city. To deepen the urban composition thus amounts to integrating two aspects that are in many ways antinomic (aesthetic versus societal), taking us back to the history of the city and the social and economic conditions of its production.

2.2.3.2. *History, mode of production and urban composition*

The understanding and knowledge of the city are inseparable from the historical field from which it is born: the city is the product of history and is born with it[21], and the real city remains deeply rooted in time and space. It therefore seems difficult, if not impossible, to talk about the city without talking about its history and the history that gave birth to it. Accordingly, if the city appears as a determined social product, it can only be thus within the framework and historical conditions that saw it emerge and develop (Braudel 1979). As such, it is also the expression and manifestation of the social relationships that characterize the society within which it evolves and develops. The city is therefore the material envelope of things and people that then takes on a meaningful form that owes nothing to chance, namely,

21. The city can also be conceived as an "abstract" form, for example, in the case of the ideal city or "utopian city". However, this is a model of the city cut off from any real and therefore any historical form.

for example, the ramparts, the winding or orthogonal streets, the different symbols of power, etc. The city in its historical configuration is part of the long term, well beyond the period that saw its birth. The city can be linked to a process of accumulation, which, when laid out, makes it possible to trace vestiges of the past. Moreover, if we consider that these different urban forms have a life span far beyond the period that produced them, the process of accumulation (to which we will return) can be examined by studying the different sedimented layers belonging to different modes of production: therefore, while the ancient stage of the city has become rare, the strata of the medieval and modern city are well and truly visible, and in many places even cultivated. This phenomenon is of course accentuated by the evolution of the city itself, as well as by the transformations that occur in the social division of labor, in the relationship between town and country, in new iterations of the state, etc. These multiple transformations can take on a local dimension. Likewise, these multiple transformations can take on different local forms, reflecting the different forms cities take here and there. That said, with the generalization of commodity production and the subsequent domination by capitalism, which has penetrated all spheres of society, we are witnessing a form of standardization in the appearance of cities, from East to West, from North to South. The city undergoes directly and/or indirectly the determining influence of the mode of production that contains it, even though it is the pure product of it. Thus, we can speak of a dialectic between the city and the mode of production, either in the form of the city as a mode of production or the mode of production by the city.

2.2.3.2.1. The city of a mode of production

A mode of production not only refers to the nature of the social relations that characterize it, as well as to the development of the productive forces, that is, the state of scientific and technical development[22]. It therefore indicates the conditions in which goods and services are produced, distributed and consumed according to a unique logic[23] and subject to historical interpretations. More generally, modes of production have direct effects on

22. For the moment, we are not establishing a difference between science and technology, but we will see later that this distinction can turn into an opposition due to the specific status that is granted to one and the singular importance that this imposes in the organization of a society.

23. Thus, for example, the mode of production of goods and services, their distribution and their consumption during the Middle Ages differ radically from the conditions of production, distribution and consumption that we experienced in the second half of the 20th century: it is no coincidence that "large distribution" did not exist between the 10th and 15th centuries.

the structuring of space, both urban and rural (even if these need to be qualified), and the whole is directly related to the nature of the production relationships specific to the mode of production concerned and the condition of the technologies in effect. Therefore, the cities of the 19th century would not have taken on their particular aspect without the prior generalization of a proletariat composed of former peasants, driven off their land and placed into salaried labor. Each mode of production has an urban expression, an urban form that appears in a more or less pure way, but which in all cases, is identifiable. It is therefore difficult to find any vestiges of the Middle Age city in the working-class neighborhoods of the suburbs built in the 1960s.

2.2.3.2.2. The mode of production of the city

As a product, the result of a production that obeys a process of implementation from the built environment, in the physical sense of the term (which does not mean that we are ignoring here the importance of the built environment in terms of the habitat, that is, of the conditions of living), the city can only ever be the expression of the material conditions that allowed for its emergence. Even if the city cannot be considered as a commodity in its own right, like others, it is no less a testimony to the characteristics of the mode of production that contains it, notably because of the technologies employed.

However, the vestiges of an earlier organization can persist over time: while capitalist forms of production have largely taken over in the economic and social organization, the ancient techniques of craftsmen of the Middle Ages still continue to produce, according to methods inherited from an earlier period, all of which contribute more or less, depending on the historical period, to the composition of the city. We find there the different historical strata which overlap and coexist in the time and space of the city.

Finally, the town/country relationship based on a specific division of labor specific to each mode of production takes on a particular social meaning (and not geographically or morphologically). It contains within it a gradation of social classes as well as the different cultural sediments that characterize it, modes of being and thinking, and any ideological upheavals it went through. The evolution of the countryside, the destruction of it by submitting to an industrial order, which is by nature foreign to it, allows us to put into context the importance of the mode of production and the concept of division of labor that each mode of production conveys.

At this point, a question arises: what are the singular characteristics of any mode of production that have an impact on the different evolutions of the city? In the context of this contribution, it is only possible to answer if we set some limits to the historical statements that underlie the question. We will stick to the characteristics of the capitalist mode of production, on the one hand because it concerns us directly and on the other hand because it presents itself through all its concrete determinants. Therefore, the accumulation of capital and the different forms that it will take during its history, constitute, in our eyes, an experience rich in meaning.

2.3. Forms of accumulation and urban forms

The city takes on the forms of the mode of production that houses it and contributes to its emergence. This is not only a technical aspect but also a more global concept insofar as it fits into the prevailing historical conditions. Therefore, if the city of the 19th century differs from that of the 20th century, then this is the result of the technical modifications that characterize each of the periods being taken into account, along with the social and economic conditions of these same periods. This being the case, the city of the 1950s, which keeps and preserves at the same time the scars of the previous periods while integrating all the movements and upheavals of the period concerned. As we have already indicated, the multiple architectural layers that constitute the city are also sociological, economical, urbanistic, etc. In the rest of this chapter, we would like to put this aspect of the city into perspective, in particular by insisting on the economic conditions in which another city emerges in connection with a new phase of ecological development. The conditions for the emergence of new forms of accumulation, that is, the growth and development of society (and thus of the city) as a whole, are thus, in our view, the privileged argument for achieving this objective.

2.3.1. *Capitalism, regulation and the regime of accumulation*

The concept of capitalism puts into perspective the interactions between the relations of production (mainly in the form of salaried labor) and the relations of competition between producers within the framework of a market economy. We now place ourselves, deliberately, within the functioning framework of the capitalist mode of production for an obvious reason: the "smart city" appears within it, after many episodes concerning the evolution of the latter, with particular regard not only to its evolution but also to its survival. One of the characteristics of this mode of production lies in its variability, which arises due to its instability, as well as in its duration, that is, its development over time. In this framework, accumulation becomes the driver behind the system's evolution. We can therefore sense the relationship between the accumulation (in all its forms) of capital and the forms that the city will have to adapt as a result of these conditions of accumulation. For example, the mass production of automobiles in the 30 years following World War II gave rise to an organization of the city that has been called the "ring city", which results from a specific form of accumulation. If we choose to adopt such a conceptual framework, this results from the difficulty, or even the impossibility, of accounting for the systemic nature of capitalism, especially as it is normatively presupposed by the orthodox approach. Indeed, whether it is a question of any institution or any exogenous factor that is likely to modify or prevent the free functioning of the market, neo-classical theory only really endorses a market free of all constraints to ensure not only general equilibrium but also its stability. Therefore, the city, in all of its concrete determinants, does not pose any more problems than the countryside, as long as the conditions of competition and the "free functioning" of markets (of currency, goods and services and labor) exist. This is especially true of a reference model that excludes any possibility of "institutional arrangements" and whose only flaw lies in the impossibility of the market behaving as it should. The city, whichever form it takes, is thus only a point in space, and its cardinal quality consists of structuring itself in accordance to the logic of the market, which remains the only possible means of regulation. As we can see, such a model (even reduced here to its simplest expression) is not able to account, on its own, for the evolutions of the city and the different forms it may have taken over the course of its history.

We have made the deliberate choice to apply the regulation approach to put into perspective the problems of the city, insofar as it seemed obvious to us that there is a direct relationship between the evolution of the city and the evolution of the economic system in which it develops. Indeed, the justification for an explicit reference to the regulation approach lies mainly in its capacity to articulate an explanation between, on the one hand, the conditions of growth, and on the other hand, the conditions for the appearance of crises that call into question this growth, and also the transition from one to the other. If we place the city and the environment at the center of our concerns – that is, the question of the sustainable city – we will easily understand the link between these two elements: Fordist growth generates a city logic that leads to the development of a "ring city"; the crisis of Fordism and the crisis of the environment are understood via the digitized city (we will come back to this idea later).

For the time being, we will retain here three essential characteristics of the regulation approach that make it possible to better understand, not only the mode of production and its functioning but also the structuring of the city that was configured by it.

2.3.1.1. *The importance of an accumulation regime*

We will first qualify an accumulation regime as a pattern or even a model for the growth of a national economy over a given historical period. More precisely:

> This term is used to designate all the regularities that ensure a general and relatively coherent progression of capital accumulation, that is to say, those that allow for the distortions and imbalances that arise permanently from the process itself to be absorbed or spread out over time (Boyer 1986, author's translation).

The concept of accumulation finds its origins in Marxist theory, mainly in the analysis of the conditions of production and the reproduction of the capitalist mode of production, whether it is "simple" or "extended"[24]. The process of accumulation puts in direct relation the coherence of the mode of production and the social relations that characterize it, while releasing contradictory forces within this process that are likely to endanger it through the emergence of deep crises. The regime of accumulation implies the idea of periodicity, which aims at highlighting the forms and specific nature of each stage in the evolution of the wage relationship, of competition, of the role of the state and of international relations. Therefore, we can put forward the conditions of reproduction and the conditions of regulation, such as the regime of accumulation and the characteristics they impart.

That said, when the factors of coherence remain dominant over a long period (here, for example, the 30 years following on from the world wars),

---

24. This theoretical set finds its main development in Capital Volume II, which, let us remember, was only published posthumously. Indeed, Marx considered these texts, published by Engels after his death, to be unfinished. As such, some have considered Engels' responsibility in the construction of "Marxisms" to be decisive.

the conditions of accumulation consist of searching for the different economic and social regularities that ensure their perpetuation, and concern factors such as the organization of production, work and the different forms that they cover, a certain sharing in the added value and the valorization of capital in time and space for the mode of production concerned. Therefore, the problematic of the regulation approach is that it reflects the procedures to be implemented in order to better understand the set of misalignments, but which are the consequences of this same accumulation and which impose others to ensure the continuity of the process.

2.3.1.2. *The role of a regulation mode*

The idea of "regulation" is borrowed from Canguilhem[25] and aims at allowing the establishment of relationships between certain rationalities (even if limited) of a systemic nature, the dysfunction of which can lead to a questioning of the whole system. Two main causes can lead to such a state of affairs:

– the various factors likely to converge towards a general equilibrium (i.e. macroeconomic) are not able to ensure it: the adjustment between aggregate demand and aggregate supply, the distribution of income and the various financial flows remain imperfect with regard to a possible adjustment;

– contrary to the orthodox neo-classical approach to general equilibrium, the interplay of institutions (to which we will return later) cannot be fundamentally consistent with the rules of pure and perfect competition. Therefore, the market can no longer function as the slight element of regulation that otherwise allows for this equilibrium to be achieved.

Following the example of Boyer, it is thus possible to qualify:

> from the mode of regulation, any set of procedures and behaviors [...] that has the triple property of (1) reproducing fundamental social relations [...], (2) supporting and steering the accumulation regimes in effect, (3) ensuring the dynamic compatibility of a set of decentralized decisions (Boyer 1986, author's translation).

25. See the author's article for greater detail: *Encyclopedia Universalis*, 15, 797–799.

2.3.1.3. *The intermediation of institutional forms*

It is clear that the regulation approach, in its fundamental theoretical formulation, is placed in the field of macroeconomics, since any mode of regulation sets up a range of institutional methods that end up shaping, even constraining, individual behavior and which thus induce a strong injunction with respect to the market mechanisms that can, therefore, no longer operate. However, the question arises as to how these institutional forms operate? If regularity is at the foundation of the accumulation regime, it is necessary to determine the active principles that make these forms the conditions for these regularities. Three factors are decisive:

– on the one hand, regulatory forms that are imposed in the form of laws and administrative regulations and which have a coercive effect on the individuals or groups to which they apply, thereby reducing their freedom[26];

– on the other hand, various elements of agent behavior, particularly during negotiations, have the effect of rigidifying offers and demands. This may also involve contractual negotiations within the context of labor relations, as well as commercial negotiations vis-à-vis prices, all of which leave little room for regulation by the market;

– finally, there are less coercive means, at least in appearance, "which aim to obtain an implicit codification and a relative homogenization of behavior […]; the community of a system of values or at least of representation", which is sufficient to modify any market logic (Boyer 1986, author's translation). Therefore, adherence to a system of norms in effect appears to be the manifestation of an individual choice, even if this choice is the result of a community or other form of membership.

Without going into the details of the regulation approach, we see that the objective remains clear: to account for economic growth and its conditions, as well as for the crisis of this same mode of development. Of course, the city does not escape this overall coherence, the regulation approach thereby allowing us to explain the transition from one urban form to another. Using a

26. For example, think of the functioning of the labor market. The orthodox model tells us that any relative variation in supply and demand on the part of companies and employees will result in a variation in the wage level, which will adjust supply and demand. However, the existence of a minimum wage stops this mechanism. The market is no longer able to adjust the quantities concerned. This reasoning can be adapted to all regulatory situations that are likely to prevent market flexibility. Hence, most orthodox liberals are calling for the abolition of the minimum wage.

distinctive criterion, based on the different forms of accumulation that capitalism has undergone over the past 70 years, we propose here to associate these forms of accumulation with the different forms that the city has taken, following the example of the relationship that we established above: between the 30 post-world-war years and the intensive accumulation that characterized this period.

2.3.2. *Fordist accumulation and urban growth*

In this respect, we will distinguish two specific periods with particular forms of accumulation, with two specific modes of regulation, each of which ended in crises, thereby explaining the need to move from one form to another. Moreover, the singular and general development of digital tools has made it possible, through integration, to move on to a new form of accumulation that we have termed "service accumulation". In addition, we will distinguish between two organizational forms of urban space, each depending on a singular regime of accumulation. In other words, the emergence of a particular form of urbanity is associated, to varying degrees, with its own regime of accumulation.

2.3.2.1. *Fordist accumulation, mass production and mass occupation of space: the 30 post-world-war years and the dormitory city (1945–1975)*

2.3.2.1.1. Accumulation regime and regulation

The Fordist-type accumulation that characterizes economic growth during the long period following World War II (1945–1975) replaces the previous form of accumulation otherwise described as "extensive accumulation" (Lorenzi et al. 1980) and which characterizes the first period of growth of capitalism (1850–1939)[27].

In reality, Fordist-style accumulation is based on a double compromise:

– On the one hand, as far as the productive system is concerned, the owners of capital and employees agree to share the benefits of growth, mainly through the sharing of productivity gains. Consequently, the growth

27. The periodization that we propose here relates exclusively to European countries. The precocity of the United States in this field inspired the countries of Western Europe. In this respect, the term "Fordism" applies to a form of accumulation that reflects the origins that these same forms of accumulation may have taken.

in income resulting from production will lead to a considerable increase in household consumption and thus in overall demand. This phenomenon is accentuated by massive investment in the productive apparatus, which will also make it possible to maintain these productivity gains throughout the period. Moreover, this strong and continuous growth is based on a standardized consumption model: mass production corresponds to mass consumption, which is all the stronger because productivity gains have led to a rapid fall in the price of the goods produced. The combination of these two phenomena, the sharing of productivity gains and a general fall in the price of manufactured goods, will lead to strong growth (e.g. in France, this was an average of 5.6% per year over this period).

– On the other hand, a new compromise is established between the state (i.e. either directly with the state or public institutions: national, regional and/or local, depending on the country) and companies: the state no longer carries the cost of sovereign activities (justice, police, army, etc.) but it participates in the general investment of equipment on behalf of the nation (social housing, social-cultural equipment, high schools, colleges, universities, health establishments, etc.). In addition to these investments, there is the establishment of a (relatively) strong social protection system: social security.

The reasons for growth boil down to two essential factors: on the one hand, the existence of strong productivity gains in industry and their redistribution within the framework of a wage policy that is favorable to wages, and on the other hand, the definition of a consumption norm that is facilitated by the sharing of added value. Moreover, the omnipresence of the state as a regulator makes it, as an actor, a fundamentally regulatory interventionist state.

The causes of the crisis lie in the reasons for growth and are consubstantial with it, namely, the exhaustion of the conditions that were at the origin of growth, associated with two other factors (which we will revert to shortly); and the weight of the state and tertiarization, the importance of which we will appreciate in the third point, concerning "service accumulation". Without going into detail, we will retain several causes for the crises of the 1970s[28], even if they are of unequal importance in triggering it[29].

28. In this respect, we list here the best-known bibliographic references: Aglietta (1976), Boyer (1979), Lipietz (1979), Lorenzi et al. (1980). The dominant evocation that oil was the

The exhaustion of productivity gains

Between 1965 and 1975, there was a reversal in the trend of productivity gains: they went from increasing to stagnating, before decreasing, the main cause stemming from what has been called "the labor crisis".

The exhaustion of the consumption norm

Such a norm, together with its expansion, played an essential role in post-war economic development. Because intensive accumulation is fundamentally based on both productivity gains and, perhaps above all, on mass consumption: in order to absorb ever-increasing production, it is essential to sustain demand, whether it comes from the private or the public sector. We are also witnessing a certain saturation of needs in the area of production of durable goods (automobiles, household equipment, etc.), which have also played an important role in growth. Finally, with the 1970s, a double reason fundamentally calls into question intensive accumulation.

Stateization and tertiarization: without going into detail, several elements make it possible to underline the importance of these two factors

On the one hand, in France, for example, the state played a decisive role throughout the post-war period: directly and/or indirectly, the state contributed 70% of GDP. At the same time, compulsory deductions rose from 41% in 1951 to 48% in 1971, at a rate of 5.6% per year. The same applies to public expenditure, which rose from 40.9% in 1947 to 49.8% in 1971. In addition, the total number of public administration staff (central and local) rose from 1,924,647 in 1958 to 3,041,715 in 1980. This development of the state gave the productive sector carried with it such financial weight that the productivity of the productive sectors was considerably affected. In this way, the state, through a growth that was poorly linked to the rest of the economy, and which in some cases was even contradictory, diverted an increasingly large share of the surplus to the detriment of accumulation.

---

cause of this crisis is, in our view, misleading, because if the very sharp increase in the price of a barrel of oil in 1973 was in fact the trigger, it was not effective enough to be the cause of the crisis itself.

29. Obviously, in our view, the reasons for the crisis and the reasons for its onset are completely different. While the surge in the price of a barrel of oil is a triggering factor, it is certainly not the cause of the crisis itself.

On the other hand, tertiarization, or the development of services, and the so-called "unproductive" employment in the sectors concerned, will play a role identical to that of the nationalization of the economy. The development of unproductive work will continue to undermine the economic surplus intended for the growth of industrial sectors. The number of jobs dedicated to tertiarization was 4,900,000 in 1955, which rose to more than 7,800,000 in 1977. In this sector, the health sector played a dominant role, which it continues to play today, and will therefore continue to weaken economic growth.

These data, as well as the mechanisms underlying them, concern the French economy, but they are symbolic of trends in the so-called developed world over this period. Therefore, what is going to be called into question is indeed the forms of accumulation.

2.3.2.1.2. The adaptation of urban forms

More specifically, the question we ask here concerns the relationship between the evolution of the city and the very nature of the forms of accumulation within which it is embedded. In other words, the evolution of the city must be assessed in terms of the conditions of accumulation that characterize the period. We mentioned above how the perception of the city corresponded to the evolution of social and economic structures, in particular by associating the technical notion of the "ring city" with a form of economic development, notably at the rate of intensive Fordist-style accumulation. Indeed, during the period in question, the rate of household car ownership increased threefold: from 21% in 1954 to 35.4% in 1962, 53.3% in 1968, 67% in 1978 and going on to exceed 80% in the 2000s. Without going into detail, 55% of motorized households have only one car, while 45% have two cars. That being said, the share of single-car households is declining, specifically, 67% in 1982 versus 55% in 2014.

It is therefore easy to understand that the conception of the city has been profoundly modified and, under these conditions, a technical response to a technical problem has been provided through the technical concept of the "ring road city". Obviously, the notion of a sustainable city is, at this point, totally foreign to the concerns of urban planners, whose objective is only to regulate the incessant flow of cars that flood the city. It is therefore easy to understand why the technical name of the city reflects the specificity of the accumulation conditions of the period under consideration. In some cases, the management of these flows is totally subject to the logic of the "all car" approach, which, through the construction of parking lots[30] in city centers, aims to bring cars into the heart of the cities themselves.

However, we could be criticized for the fact that there is a relatively long gap between the period that characterizes this form of accumulation and the adaptative conditions of the city to the new conditions that have been imposed by these forms of accumulation. This argument does not seem to be acceptable to us for two essential reasons: on the one hand, the rhythms of accumulation must take into account the conditions of adaptation for all sectors of society. In other words, the movement of infrastructures to adapt to the new conditions of accumulation and its consequences cannot be grasped in too short a time. On the other hand, the inertial effects of technical change imply a response time to the new conditions created by the new forms of accumulation. The ring city cannot therefore be anticipated before the socio-economic phenomenon of the car has itself developed strongly and irremediably.

2.3.2.2. *New forms of accumulation and changing urbanity: from Fordism to post-Fordism*

As we specified in the previous section, one of the reasons for the crisis of Fordism lies partly, and only partly, in the emergence of the service sector. Due to its growing importance as of the 1970s, it burdened the functioning of the economy with additional costs and thus contributed to the crisis. Western economies will therefore move from an "industrial" (or even manufacturing)[31] economy to a service economy, that is, an economy in which intangible service values become dominant with the corollary of mass unemployment. The forms of liberalization of the economy, internal and external, will at the same time allow for the globalization of trade and the global financialization of the economy. Neoliberalism and the new forms of accumulation and regulation will therefore be the result of all of these contributory elements, which, moreover, will inevitably be accompanied by changes in institutional forms.

---

30. The failure of such a strategy, which has led to massive congestion in large urban centers, seems highly significant to us and has caused a radical change, accentuated by the effects of urban pollution, which are increasingly strongly denounced for their dangers in terms of public health.

31. We share the analysis that Western economies are abandoning industries in favor of importing from China (in particular, but not only) then qualified as "intensive plowing", i.e. of a manufacturing nature (e.g. textiles), so much so that China has been called "the workshop of the world". For further reading on this subject, see du Tertre (2009).

2.3.2.2.1. Liberalization of economies as a condition for new forms of accumulation and regulation

With the arrival to power of Margaret Thatcher and Ronald Reagan in England and the United States, respectively, the neoliberal doxa spread and led to major changes in public policy. The doctrinal foundations of financial liberalization can be summarized as follows: the financial sphere can exert powerful driving effects on economic growth. To achieve this, the financial sector must free itself from the constraints imposed by state regulation. If the financial sector is deepened at the end of the process (internal liberalization), the expected benefits in the real (domestic) sphere must be significant. It then becomes possible to move on to the next stage by opening up the capital account by allowing capital to flow in and out (external liberalization).

The operating methods of internal liberalization

Generally speaking, liberal authors who thought about economic liberalization initially emphasized the banking system and its role in economic development, thus linking the importance of economic development to the development of the banking system. The deregulation of prices in general, and of interest rates in particular, and the importance of financial savings as a prerequisite for investment and thus for economic growth, gave the previously constituted financial sector a place and a determining role. Indeed, banks mobilize savings and select only profitable investment projects, which has the effect of driving growth upwards. It is therefore necessary to ensure that the market functions freely to collect these savings and to reject all forms of constraints. In this perspective, the system of compulsory reserves, credit controls or preferential interest rates linked to state industrial policies must be abandoned. In this area, the role of the state is to limit its action to regulating the money supply in order to avoid inflation.

Deregulation and external liberalization

The liberal conception asserts (without demonstration!) the benefits of international capital mobility. This allows countries with low savings rates to attract financing for domestic investments. Moreover, international investors can diversify their portfolios, thereby spreading the risk. Removing all restrictions on capital movements, that is, by preventing capital from flowing in and out unchecked and unhindered, is what the liberals call deregulating capital movements. In this respect, according to neo-classicists, there are many arguments: deregulation favors a more efficient allocation of savings, allowing greater possibilities for diversifying the risks associated with investment, and thus encouraging faster growth and the attenuation of business cycles.

2.3.2.2.2. Post-Fordist accumulation: financialization and globalization

The post-Fordist accumulation regime can be characterized into two ways:

– On the one hand, by a structural transformation of the production of value, as well as by a change of form in the appropriation of value. Indeed, we are moving from what has been called the "industrial model" to a "service model". This means that the productive objective initially resided in the design of the manufacturing and marketing of goods. Therefore, within the framework of such a model, the structural division of the economy as a whole obeyed the following breakdown: agriculture, industry and trade (or services), with the overall supply being limited to the activity of these three major economic sectors. We can summarize the changes that this structural transformation has imposed on the economy as follows: saturation of the markets for durable goods consumed by households; a change in production structures in favor of the so-called "service" sector[32]; the dominant character of intangible outputs and finally, the new information and communication technologies that have largely contributed to modifying communications and therefore the systems for locating activities. In addition, the new environmental requirements have imposed the implementation of regulatory measures with regard to the emerging economic dynamics.

– On the other hand, through an international division of labor in the production process. This internationalization is positioned at three distinct levels. On the one hand, in the field of capital, the internationalization of finance leads to the structuring of a sector in its own right, with its own markets, its own players, its own rules, as well as to profound changes in the functioning of companies in all other sectors, whether public or private. Therefore, it is easier and faster to raise capital, and investors take advantage of this opportunity to ensure short-term profitability that is a radical departure from the patrinomal vision of a company. On the other hand, a

32. The development of the service sector (see below) has been to the detriment of the industrial sector, mainly in the so-called "labor-intensive" sector, where most activities have been relocated to areas where labor costs are far lower than in the country of origin. This explains, for example, the virtual disappearance of the textile manufacturing sector in France, which was considered too "labor-intensive" and hence relocated to Southeast Asian countries.

second level comes into play, notably with regard to consumption. Indeed, most markets are not global markets but are constructed by operators on the basis of international subsets: Asia, North America, Europe, South America, Russia, etc. In any case, these subset units go far beyond the narrow framework of nation-states, which constituted the organization of markets during the Fordist period. Finally, internationalization also concerns production, notably through an international division of production between regions specializing in R&D, finance and logistics, while other regions specialize in manufacturing, maintenance and servicing, as well as regions that concentrate on the main consumer markets. Furthermore, this international specialization is constantly changing and evolving. For example, R&D activities have developed considerably in regions that used to specialize in traditional production activities (e.g. China, India, Brazil, etc.).

In this context, the competitiveness of companies relies mainly on innovation and know-how, in terms of organization and cost optimization, notably through outsourcing. In the so-called "mature" consumer markets, the development of companies is based on their ability to create new needs and to remain the sole supplier for as long as possible. This niche strategy is obviously subject to competition, and moreover, these markets remain small. This situation gives rise to partnerships between leading and innovative companies in their chosen field. Therefore, the logic of integration into the international economy persists in the post-Fordist accumulation regime; however, it is based on the search to minimize financial risks rather than on the desire to control all stages of the production process, as in the Fordist regime. The result is the structuring of financial holding companies that take stakes in competitors as well as in other sectors of economic activity in order to minimize risks: by spreading the assets, the risks taken are spread over time and the production space.

2.3.3. *Digitization and service accumulation*

The evolution of the conditions for accumulation at the dawn of the 21st century can be summarized around two determining elements: on the one hand, the irruption of information and communication technologies, which, because of their sudden importance (something that was even violent for some), imposes a unique form of accumulation; on the other hand, the importance of the service sector, which gives the forms of accumulation a particular trait. We will briefly present the two characteristics of this new form of accumulation as well as the qualitative effects of these two specific modalities of post-Fordist accumulation.

2.3.3.1. *The emergence of a service system*

If we stick to a simple quantitative observation, it seems that the evolution in the structure of the activities which participate in the production of value (here, the GDP) is particularly significant: during the first decade of the 21st century, services represent 75% of this national activity, particularly in economically developed countries. This restructuring was largely to the detriment of the two other sectors, agricultural and industrial, whose decline in the creation of value is clear and unambiguous, but more so for the agricultural sector. Our objective here is not to analyze the reasons and causes of these structural changes in the production of value in detail, a few figures are sufficient, in our opinion, to show the extent of the phenomenon so as to place it in a systemic logic that gives it meaning. However, we will make two remarks:

– From a qualitative point of view, when we refer to the production of tangible goods, we take into consideration the material forms that circulate in the forms of purchases/sales, in a way that is physically independent of the producer and the consumer. Conversely, in the case of the production of services, if there is indeed a material form[33], it does not allow for a relative independence of the products. This situation becomes particularly evident in the context of international trade relations: the goods exchanged have crossed the border and are sold to users who are often thousands of kilometers from their place of production, whereas, to consume a medical service interactively, for example, the person who wishes to consume the medical service must travel[34].

– From a quantitative point of view, particularly in terms of the production of added value and the contribution of each of the sectors to its formation (and thus to GDP), the relative weight of the services sector and its evolution within the framework of the rhythms of accumulation as we have known them has changed considerably. To us, two sets of figures seem to be particularly explicit in this area. On the one hand, from the point of view of the contribution to the creation of added value (i.e. GDP), a

33. Therefore, the individual who leaves a hairdresser's salon is not quite the same as the individual who entered it a few minutes before.

34. Even though the implementation of information and communication technologies tends to modify this approach somewhat, this in no way modifies the notion of "service" as we use it here, as well as its implementation and conceptual application.

comparison over time demonstrates this significance: in 1962, agriculture contributed 21% to the formation of GDP, compared to 37% for industry and 42% for the tertiary sector (i.e. the service sector). In 2017, the tertiary sector accounted for 75.9%, industry 20.3% and agriculture only 3.1%. If we translate this into terms of employment, the figures speak for themselves: in 1954, agriculture contributed 26.6% of overall employment in France, 35% for industry and 38.4% for the service sector. In 1980, the figures were 8.4%, 35.1% and 56.4%, respectively. In the first decade of the 21st century, these figures average out at 3.9%, 21.9% and 74.2%. This massive phenomenon is reflected in the structural modification of cities, regardless of their size: the entrance to cities has undergone aesthetic changes that everyone denounces, in particular the disfigurement of the immediate periphery through anarchic development of so-called "commercial" areas whose primary activity is typically reserved for services[35].

The rise of services, particularly intangible production in general, throughout the economy, has been facilitated by digitization, and has been accompanied by a certain change in the way they are thought of, produced and consumed. Cities have the opportunity to seize this transformation in service of new arrangements between governance, "common" services and collaborative practices.

In fact, during the last two decades, the mode of accumulation has undergone some structural transformations that substantially change the conditions of production.

Market saturation for household capital goods (see paragraph 1 on the exhaustion of the consumption norm), a result of the saturation of the consumption norm following on from the Fordist era, has changed the conditions of world competition and globalization. As a result, productive structures have been transformed to the exclusive benefit of the service sector and the production of intangible goods[36].

In addition, the production process integrates within it an increasingly important share of intangible inputs, which have as a result become dominant. In this respect, information and communication technologies have

35. For further reading, see the excellent contribution on this topic by Garcez, C. and Mangin, D. (eds) (2014). *Du Far West à la ville : l'urbanisme commercial en question*. Éditions Parenthèses, Marseille.

36. The "uberization" phenomenon of society is the most convincing indication of this mechanism, affecting so-called developed societies in particular. We will come back to this.

considerably modified the conditions of communication and thus localization conditions (and delocalization).

Finally, even if this remains marginal, environmental requirements refer us to the relevance of the multiple mechanisms for regulating economic dynamics.

Clearly, the developments in structural terms leave no doubt as to the dominant character of the service system, which thus leaves a definitive imprint on the characterization of the forms of accumulation. If we relate these evolutions to the configuration of the city and the aesthetics that emerge, one observation becomes evident: the entrances to all our cities, large and medium-sized (and more!), expand as far as the eye can see, an infinite quantity of "superstores" dedicated to commerce and services, and generally replacing manufacturing activities that have been "relocated", generally to Asia. In some circumstances, the old local factories that were the pride and joy of producers have been transformed into "cultural" places or else into giant "sales rooms" that are dedicated to commerce. As an example, take the old wool factories of Roubaix, which have become gigantic tertiary venues dedicated to trade and services. This illustrates how the form of accumulation shapes the forms that cities will take as they move into the future.

2.3.3.2. *The role of information and communication technologies*

The rapid and profound irruption and the irreversible penetration of information and communication technology within society as a whole are also at the core of a number of important transformations of the conditions for accumulation. The transition to digital platforms as a means of conveying information is clearly at the heart of this evolution, which we summarize as follows:

– on the one hand, they make it possible to considerably increase the information processing power, thus facilitating the solution of complex problems[37];

– on the other hand, and as a consequence, they facilitate the human/machine interface that allows operators to take an interest in the regulation processes;

37. Moreover, the new processing power of computers is at the origin of what is being called "artificial intelligence" whose existence (real and/or potential) is tied to enormous processing power and vast information storage capacities, a feat which humans are not capable of.

– finally, they encourage the renewal of the conditions for the dissemination of information, which, thanks to high-speed levels, have encouraged its uptake by all actors, wherever they are based and whatever their work entails.

If the effects of the introduction of New Information and Communication Technologies (henceforth NICTs) are comparable to the effects of introducing new technologies in general, which characterize capitalism, it is worth noting two specificities which have had a heavy impact on actor performances. On the one hand, with NICTs, the possibility of processing superabundant amounts of information and data becomes possible and is even part of the structural logic that this introduction implies. On the other hand, they make possible the emergence of a form of permeability between the professional activities of agents and their personal activities outside of work[38].

However, an overly mechanical approach to the notion of information leads to ignoring certain aspects of the consequences from the introduction of NICTs. On the one hand, this leads to an overestimation of the technology itself, which is then given qualities that it does not have; on the other hand, there may be some confusion of a semantic nature, between what is in the domain of data, the domain of information and the domain of knowledge and knowledge formation. This is because data and information can have a meaning that the digital equipment is not able to know and recognize. Finally, what we mean by knowledge is a matter for science (and not technology) and therefore needs more complex systems.

As we saw in the previous paragraph, consumption is oriented towards capital goods, which are either renewal or multi-equipment. Most markets for manufactured goods are thus relatively saturated and slow-growing. The markets for ICT products are no different, and their diffusion also experiences the same problem. As a result, innovation processes remain at the heart of the entrepreneurial dynamic, without functioning as a relay for growth. Moreover, consumers are increasingly volatile; as a result, the radical uncertainty of the environment destabilizes the foundations of

38. The economic crisis and its consequences on employment as a result of the Covid-19 pandemic is a perfect illustration of this: due to the digitization of workstations, the response to this new situation has been mainly to move the workplace from the "office" to the home. This possibility, which was quickly exploited, reflects the reality of this permeability, which has gone from theory to very real. The emergence of telework would not be possible without ICTs.

investment amortization and therefore affects the profitability of capital. Several observations:

– The so-called "service economy": the service economy has become dominant in contemporary economies insofar as more than 70% of the added value (i.e. of the GDP) as well as employment is due to it. The main differences with the industrial economy can be summarized as follows:

- services are not stocked; which on the one hand avoids a cost, but on the other hand does not facilitate adjustments between supply and demand, the pace of innovation being set through the emergence of new demands,

- services are provided in "real time" and places the customer at the center of the relationship, which can strain the wage relationship,

- the development of services not only depends on solvent demand but also on factors linked to information and communication.

– The generalization of information technologies: the transition to digital technology as the main vector for the circulation of information has facilitated three things: on the one hand, increasing information processing power to previously unknown magnitudes on the pretext that situations are increasingly complex; on the other hand, developing "human–machine interfaces" that facilitate the implementation of regulation mechanisms; finally, mastering the conditions for information dissemination facilitates accessibility to a large number of actors. However, to date, a simplistic idea on the notion of information tends to prevail: information can be formalized and integrated into a digital system, and the company would add further value if it were able to set up expert systems, in other words, a "knowledge economy". However, the "cognitive capital" approach poses certain problems: first of all, there is an obvious confusion between on the one hand data, information and information processing, and on the other hand, knowledge. Here, we find the relationship between science and technology, particularly in the relationship between the question of "why?" and the question of "how?". This is because science answers the questions as to "why?", whereas technology answers the question of "how?". We are therefore dealing with two fundamentally different issues that complement more than they oppose each other. Secondly, over time, we are witnessing an overestimation of existing (or future) technical systems of coordination systems between actors. The "data" are elements that appear in measurable, quantifiable and integrable forms in computers that are capable of doing everything except grasping meaning. Finally, knowledge carries meaning and is therefore more complex and must be integrated into systems of thought, the interpretation of which requires a heuristic dimension on the part of the technologies that are incapable of making them intelligible. Therefore ICT, while it allows productivity to be improved, generates wasted time through the production detours that it imposes, particularly in the selection of relevant information: what is gained on one side is lost on the other. The answer to this major problem does not lie in the technology but in the space of knowledge, that is, in the possibility of matching the different contents of this knowledge. The subjectivity of the observer within the technical space becomes decisive. Finally, a last problem arises in the diffusion of ICTs: they result from the permeability that appears between work time and non-work time. For, if we can take our work outside the workplace, this implies that the workplace can correspond to the private sphere. The notion of work thus becomes more abstract, less understandable, and therefore presupposes an institutional response to regulate its intensity and meaning.

– The role of negative externalities: the intensive pace of growth in manufacturing activities has led to an exponential increase in negative environmental externalities, thus calling into question the renewal of natural resources that are essential to life[39]. Generally speaking, it is society's relationship with nature that is affected, by calling into question the renewal of the primary resources that are essential to human beings, so much so that companies seem to be increasingly concerned about this aspect of the process of production and accumulation, and thus to reconsider their responsibilities. We are therefore witnessing a form of regulation, albeit very modest, which involves new actor dynamics, particularly in the context of sustainable development. These are new systems of responsibility that result in debates on quality, traceability and "corporate social responsibility". The legitimacy of the resulting procedures thus depends on the mode of involvement of external and internal actors in the entrepreneurial system. Provided we continue as per the results obtained to date, it is clear that little has changed.

– Economic performance and "service" accumulation: in reality, service accumulation can be deduced from the inefficiency of the post-Fordist accumulation approach that focused on the uncertain nature of the short-term context sought by the neo-Fordist and financialized enterprise. Admittedly, it is too early to characterize a successful model for a service-type entrepreneurial system; however, it is possible to observe that some

39. On August 18, 2016, humanity had exhausted its share of known natural resources for the period, with the balance taken from the remainder of the current year drawing on the following year. In 2017, this date occurred on August 4. This is how quickly we mortgage the future of future generations.

companies modify their approach, and through this, it is possible to identify a certain number of avenues, namely:

- the shift from "short-termist" to long term in the evaluation of performance indicators for the accumulation process itself,

- the inclusion of a fourth factor in this evaluation, in addition to quality, productivity and profitability, that is, externalities,

- the establishment of indicators that are distinct from performance indicators and associated with the heteronomous characteristics of the companies that drive the accumulation process.

2.3.3.3. *The city under the service accumulation regime: the smart city*

As we saw earlier, service accumulation results from a transformation of economic activity (mainly from a functional point of view and not from a structural point of view) by affirming the primacy of new information and communication technologies, which are being made available to a service society for which they will become the core element. Consequently, we can logically think that we will witness a transformation of the city which, with these new conditions, adapts to the new forms and modalities of accumulation: just as Fordist accumulation led to a city that was constrained by the massive development through vehicle usage, ICTs impose new forms and modalities of identification onto the new city. This is due to the fact that not only has the city become "smart" but also the habitat itself is adorned with the ornaments of this intelligence. We will highlight two essential aspects.

On the one hand, the particular meaning that we give to the smart city, which we can summarize as follows: the city's intelligence character corresponds to the elements of the city, which gives it a certain aptitude for satisfying the requirements for improving the quality of urban services, namely, the economic optimization for the use of ICTs, a form of social equity in the use of ICTs (openness, adaptation and sharing, institutional relevance in the use of ICTs) and finally, an environmental control of all components that use ICTs. As we can see, the city is in fact becoming intelligent due to the explicit references of sustainable development criteria.

On the other hand, the assimilation between the smart city and sustainable city seems obvious insofar as the state of the planet does not allow us to dissociate the two qualifiers as this would be antinomic. The new definition of the smart city thus becomes: the quality of "smart" and "sustainable" developments impacting the city that correspond to the characteristics of the latter, which therefore gives it the capacity to satisfy the system in terms of the requirement that can be summarized as follows:

– sustainability of the economic control of its development and economic optimization in the use of ICTs;

– social equity in economic, social and cultural development in the use of ICTs;

– openness, sharing and adaptation in the social use of ICTs;

– environmental control of all the components likely to develop, as well as all the elements constituting ICT uses.

Therefore, economic, societal (social and cultural) and environmental requirements become the reference requirement system to define and characterize the smart city. We can only note the sum of difficulties that appear insurmountable when we consider the means that would have to be implemented in order for cities such as Sao Paulo, New Delhi, Mumbai, Beijing and the many other megacities of a similar nature to become smart cities, in the same manner as defined above. A truly enormous mission! We propose to further analyze the considerable impact that this urban movement has had, not only on the immediate lives of the people concerned but also the particularly negative impact on the evolution of the planet, in many areas, but above all from the ecological standpoint.

Nevertheless, this observation, however severe it may be, should not prevent us from suggesting, as well as from thinking about putting in place strategies that must be part of safeguarding an urbanity that has become human, an urbanity that would make the city resilient, a city whose structures would be able to meet the challenges while respecting the environment and safeguarding human pursuits. In the following sections we propose to address the problem of the future city, a city that is capable of surviving on its own, a city that is as resilient as it is sustainable, capable of taking up the challenges that arise… if indeed such a thing is even possible.

2.4. Urban space, environment and resilience

Throughout the *Trente Glorieuses*, the 30 years that followed World War II, in addition to the major changes that accompanied the strong economic growth characterizing this period, the city and urbanization underwent profound transformations linked to this specific mode of accumulation: an increase in the size of cities resulting from demographic pressure on the one hand, as well as from the relative and continuous abandonment of the countryside (rural exodus) on the other, property development and rental construction, and also of the mode of occupation of space due to immigration and the generalization of vehicle usage. In short, the city expanded at the same pace as economic development, while at the same time leaving behind deteriorating urban living conditions; the quantitative largely taking precedence over the qualitative[40]. We are thus witnessing an explosion of cities throughout the world, both in developed and developing countries; moreover, the mapping of conurbations led to the definition of what are now called "megalopolis" (which differ from "megacities" in terms of the quantitative and qualitative amplitude of the population and the space occupied), whose populations are counted in the tens of millions[41]. The operations to expand city limits thus accelerated throughout this era, while the overall organization of the urban system was not rethought: urban spaces, roads, rural routes, expressways, that is, a set of discontinuities end up imposing themselves progressively and irremediably. According to Paquot, new cities increasingly belong to what he terms "urban disasters", lacking a clear and distinct desire to think towards the habitat quality of future occupants.

Faced with this relative failure of individual and collective housing, based partly, but only partly, on land issues, we also note the absence of

40. The importance of this expansion phenomenon of urban areas can be appreciated by measuring the extent of city expansion into the countryside: over this period, it is considered that every 10 years useful agricultural area (UAA) lost the equivalent of two French departments to the benefit of the cities that were expanding. For further reading on this subject, see Paquot (2015).

41. While a megacity is a huge city in terms of size and its number of inhabitants (e.g. Tokyo, Sao Paulo, etc.), a megalopolis is made up of a set of cities linked by network systems (communication routes, roads, trains, canals, as well as intangible services), which support the exchange of people and/or goods. These megalopolises form a very densely populated whole and integrate a large number of economic activities. There are currently three megalopolises in the world, even if some authors have redrawn their features in recent decades. In the United States, there is a zone that includes megacities such as New York and Boston, and goes from Washington in the south to Boston in the north, called "BosWas". In Japan, there is a megalopolis that includes Tokyo, Osaka and Kobe. Finally, in Europe, there is the "European backbone" that amalgamates the cities between London and Milan, including Rotterdam, Brussels, Paris, Frankfurt and Zurich. For more on this topic, we recommend Gottman (1961) and the more recent work by Le Diascorn (2006).

appropriable spaces, whether individual and/or collective: the large housing estates leave nothing for the inhabitants to use, reducing housing to what has been called "dormitory towns". From there, the idea developed that the solution lies in the individual house, which thus appears to be an alternative solution without taking into account the long-term infrastructure and management costs that this type of (false) solution imposes. Moreover, this peripheral development of the city will be reinforced by the massive and irreversible development of the private car, the general uptake of which was made possible by the strong growth of "intensive" accumulation over this period. The emergence of the suburbs is the result of the conjunction of these two phenomena, on the one hand, the multiplication of low-cost housing outside cities and, on the other hand, the massive proliferation of private vehicles, which became all the more necessary as the distance from the work centers increased with the continued urban growth. With the 1980s and the prolongation of the economic crisis, considered to be a "profound crisis", this process was not modified, even continuing to grow, and the neoliberal model of accumulation that followed only reinforced this trend. In France, the marginalization of the "*banlieues*" ultimately led to create a specialized "ministry" specifically dedicated to solving problems associated with these parts of the city[42].

The question that arises from this observation can be summarized as follows: is the tremendous urban development that the world economy is experiencing compatible with harmonious development for all humanity? In other words, is the evolution of the "world economy" and its consequences for the organization of the city compatible? Can the dangers posed to the environmental future of the planet, particularly in the context of the so-called "Fordist" and "post-Fordist" cities, be controlled within a framework promoting the implementation of sustainable development, the standards and constraints of which would be applied to the city? The emergence of complex urban systems and the resulting risks, which increase the possibility of all kinds of disasters, are all factors that transform the perception of the secure and hierarchical, but stable, city into a city subject to the devastating risks of all possible disruptions. The resilient city thus becomes a new mode of management for this complexity and leads us to perceive it as a system capable of adapting, of recovering from the difficulties that are likely to endanger it. The sustainable city and the resilient city therefore need to co-opt the other, to develop jointly in order to ensure a certain continuity of

42. Remember this creation is the work of Mr. F. Mitterrand and the government of Mr. Bérégovoy in April 1992 and that Bernard Tapie was the first incumbent.

both, of one through the other. The question thus becomes: can cities such as Mumbai, Sao Paulo, Lagos and the many other sprawling cities with galloping demographics, which have already damaged their environments so much, be the object of resilience such that their future (and that of humanity with it) is guaranteed?

2.4.1. *Demography and urbanization: a major challenge for the environment*

Urban growth as we have known it over the past 120 years is actually facing a dual problem: on the one hand, there has been an unprecedented exponential increase in demography at the global level; on the other hand, a strong urbanization[43], which is as much the result of the galloping demography as from the attraction of cities and their alleged benefits, typically to the detriment of the countryside.

2.4.1.1. *The weight of demography and the city in the making*

Indeed, since the beginning of the 20th century, the world's population has been growing rapidly: 1.5 billion people in 1900, 2.5 billion in 1950 and more than 7 billion in 2010[44]. We are faced with an exponential growth of the world population, a trend that, if it were to continue over time, will cause problems that would quickly become insoluble. With economic development and its corollaries, improvements in general health, standard of living, a significant increase in life expectancy (45 years in 1950, 70 years in 2016) and a decrease in mortality, world population growth remains constant, even if it is tending to decrease, particularly with the drop in the fertility rate from five children per woman in 1950 to a little over two in 2016.

At this point, however, a question: is such a population growth rate (even if we are witnessing a decline in rates) feasible, that is, sustainable over time and under what conditions? The forecasts of the UN demographers who produced the figures quoted above envisage a world population level that is

43. We define the urban as a territory within which the building is continuous, that is, where there is less than 200 meters between constructions and which brings together a minimum of 2,000 people.

44. The world population should reach 9 billion in 2050, of which nearly 50% will choose to live in cities. This population growth rate, if it turns out to be correct, should first impact the developing and "emerging" countries.

close to 11 billion human beings by 2100. However, if we take into account the major trends that are already emerging, these evaluations (by the same UN demographers) do not lead to this growth rate of the world population, but conversely to a decrease in it, that is, to a decrease in the absolute decline of the world population. For Bricke and Ibbitron (2020), there is a low range of the scenario envisaged by this same organization at 7 billion people, that is, less than today. The causes of such a decline would be structural and would lead to a drastic decrease in births on all continents, without exception: urbanization, distancing of family ties, decline in religious sentiment and its societal impact, increase in women's autonomy, etc. are all factors that would tend to amplify a phenomenon that is already observed in countries such as Japan, South Korea, Italy and Ireland. This means that countries that wish to maintain their population levels will have to resort to immigration. However, even migrants are getting older and their fertility rate will also be falling in the relatively near future.

That said, the current urban growth as we have known it, and as we know it, has had an immediate consequence to the explosion of certain cities: in 1950, one person out of three lived in a city; in 2015, this increased to every second person living in an urban setting. Even if the growth rate of cities decreases relatively, projections established in 2050 predict that 80% of the world's population will live in cities (this figure is independent of the amount of world population at that time: whether it is 11 or 7 billion). For example, the galloping forms of urbanization as observed in China, India (Beijing, Shanghai, Mumbai, New Delhi, etc.), Latin America (Sao Paulo, Rio de Janeiro, Buenos Aires, etc.) or Africa (Lagos, Kinshasa, etc.) only confirm, for the time being, the growing mass of cities to the detriment of the countryside. This last remark does not fail to pose a problem insofar as, for the time being, only the countryside is able to feed the city.

The consequences of such a situation are obviously particularly worrying for the environment and for the future of the planet: urban expansion leads to an artificialization of the soil, the consequences of which we are only now measuring, fossil fuel infrastructures are likewise exploding, the effects of urban heat islands and water management are becoming problematic, as well as compromising the future through global warming, to mention only the immediate effects of such a process on a global scale. In reality, the multiple attacks on the environment caused by the cumulative phenomena of economic growth, which is the product of accumulation as described above, and the increasing consumption of products derived from fossil fuels, find in urbanization and its consequences an unfortunate outcome that could endanger the planet.

What we have just described in terms of urban development of the "world economy" is just as valid for France in particular, and the countries of Europe in general. We generally find the same trends in the relatively homogeneous framework of developed countries. Indeed, we find that the general trends and the model described above also apply, even if the specificities of each only appear here and there. If the forms can be modified, the fact remains that the process remains the same. In terms of both the demographic aspect[45] and the environmental approach, the different urban situations seem to be "double-sided", modeled on a general model of growth that is more or less identical (excluding unique local sensitivities), which can be summed up as follows: all the countries that initially experienced strong urban growth as a result of various factors that can be found almost everywhere, namely, the attractiveness of the city and urban life, the search for a stable job and steady remuneration, a rurality with fewer attractions thereby causing a relative desertification of the countryside and, finally, a stabilization of the urban growth and its population. This pattern, even if it appears a little too "classic", is nevertheless true almost everywhere in the world, although certainly at different periods in history, including now in developing countries or "emerging countries". The tables presented below perfectly illustrate this phenomenon, the fundamental modification occurring in the ranking of the top 10 most important agglomerations in the world, between 1975 (Table 2.1) and 2019 (Table 2.2), which is, in our opinion, an illustration of the phenomenon described above.

In 1975, there were seven cities with more than 10 million inhabitants, four of which were located in developed countries. We note, moreover, that in comparison with earlier dates (1950), London had the same population in 1975, while at the same time the populations of Sao Paulo and Mexico City

45. Although, as far as France is concerned, in comparison with other European countries with substantively equivalent socio-economic levels, the birth rate remains relatively higher. Indeed, the rate remained above 2 for a very long time, that is, above a normal reproduction rate and only started to fall below this limit after 2017. Therefore, countries with high rates of population turnover, such as Ireland and Italy, experienced higher birth rates before the onset of only two per family.

had increased fourfold. However, the vitality of the urbanization phenomenon does not diminish (in relative terms) because a city like Los Angeles can increase from 4 million inhabitants to 10.8, and Paris from 5 to 9.2 million. We also note the beginning of the emerging phenomenon of urban growth in certain developing countries, particularly in Asia.

Cities	Population (million inhabitants)
New York	19.7
Tokyo	17.7
Shanghai	12.4
Mexico	11.9
Los Angeles	10.8
Sao Paulo	10.7
London	10.3
Peking	9.3
Rhin-Ruhr	9.2
Buenos Aires	9.2

Table 2.1. *Populations of the 10 largest cities in the world in 1975*[46]

It is with the new forms of post-Fordist accumulation and the massive development of international trade that the phenomenon of urbanization at the global level will become widespread. The globalization of the economy and the vertiginous expansion of financialization across the world will lead to a demographic explosion that will be reflected in the phenomenon of massive urbanization at the same time as the displacement of gravitational centers for functional modes of production. The following table highlights the continents where the activity is moving, which will make these new production sites the "workshops of the world".

46. Source: United Nations: Population Division (1980). Urban, rural and city population (1950–2000), as assessed in 1978. New York.

Cities	Population (million inhabitants)
Hong Kong	65. 7
Tokyo	42.7
Sao Paulo	36.3
Jakarta	35.1
Delhi	26.4
Shanghai	26.3
Seoul	25.5
Cairo	24.4
Mumbai	24.3
Manilla	24.1

Table 2.2. *Population of the 10 largest cities in the world in 2019*[47]

Following on this, cities like New York (a little more than 23 million inhabitants), Canton, Lagos and Mexico City confirm the displacement of urbanization towards areas that will experience (over the first two decades of the 21st century) a strong concentration in cities at the same time as an industrialization that will transform the Asian region into what has been coined "the workshop of the world".

At this juncture, we will make several remarks.

Under the simultaneous influence of three determining factors[48], the urbanization of the planet is growing faster than the world population. This means that the rate of population transfer from the countryside to the city is

47. Source: www.populationdata.net/palmares/villes/. However, if all the rankings offered by other sites differ from each other somewhat, the same trends are found, thereby confirming the cities presented above, as well as the same demographic conditions concerning the migrations of rural populations to cities in equal proportions.

48. These three factors are made up of, on the one hand, the natural evolution of rural population migrations to the cities, regardless of the reasons, and on the other hand, both internal and external migrations and, finally, geographic expansion due to the increase in the overall population, particularly in emerging or developing countries which are experiencing the highest rates of population growth.

high and continuous[49]. This phenomenon is characteristic of emerging countries, whose rural population is still relatively high.

The "triumphant" cities of the industrialized countries in the 1950s and 1980s gave rise to the giant cities of emerging powers, which have seen their populations explode, while at the same time a major population flow is moving from the countryside to the cities. China and India, as well as Brazil and certain African countries (Democratic Republic of Congo, Nigeria, for example) have seen their cities grow exponentially at the same time as their productive capacity.

This galloping urban growth has led to a significant increase in the size of urban areas[50], whose total surface area is expected to triple for a population that will only double. This last point, but not the least significant, refers to the phenomenon of peri-urbanization to which we will return later and whose importance tends to become problematic.

The demographic explosion is in fact occurring at the same time as the explosion in international trade, a consequence of the redistribution of the international division of labor. The massive relocations that follow this redistribution of world production (more frequently called "globalization") favors countries with "low labor costs" (whose production is described as "labor-intensive") to the detriment of countries who try to specialize in highly innovative productive systems (whose production is qualified as "capital-intensive"). This transformation of the world productive system, which began at the end of World War II, continues to the present day. The first effects of this movement of accumulation are quickly absorbed by the city, which suffers the first effects.

2.4.1.2. *Demographic expansion and peri-urbanization*

The concept (or category) of peri-urbanization seems to have the objective of going beyond the usual classification between city and country,

---

49. However, we are reasoning here on average values that exclude any form of nuance. Indeed, in 2011, 78% of the population in developed countries lived in urban areas compared to only 47% of the population in developing countries.

50. We borrow the definition of "urban areas" from INSEE: an urban area is a set of municipalities in a whole, without an enclave. This set includes a solitary urban center which offers at least 5,000 jobs, a peri-urban ring made up of rural communes (or "urban units") of which at least 40% of the resident population has a job and works in the urban area (or the municipalities that constitute them).

urban and rural, with the aim of renewing and adapting the historical, even ancestral, relationship between city and countryside. In this respect, peri-urbanization appears as an interface, an instance of interpenetration of the one by the other. It is therefore a "hybrid space", "neither city nor country, but one that is constructed through the blending of forms and processes specific to both categories, while being a category in itself, with its own characteristics" (Arnould et al. 2009). For these authors, the concept of peri-urbanization is also ambiguous because, on the one hand, it refers to forms of individual housing with an extension of subdivisions, and, on the other hand, it expresses a form of "urban sprawl induced by the development and lengthening of commuting between home and work"[51].

The phenomenon of accelerated urbanization remains dominant throughout the post-war period. Even if it takes on different forms, depending on the period and the country (in particular because of the structural differences between developed and "emerging" countries), the call of the city remains strong. The phenomenon known as "peri-urbanization" corresponds to the development of a form of individual housing in response to the deployment of collective housing in the emerging suburbs; the detached house becomes the strong symbol of this[52]. The technocratic and authoritarian urbanism of these *Trente Glorieuses* following the world wars, accompanied with the increase in private car use as a means of travel, has largely contributed to the development of the phenomenon (Jacquart 2005). At the same time, as the tower blocks and council flats of the large high densities were being built into the urban space, and the city was becoming increasingly more difficult to live in, suburban housing was developing on the outskirts of large cities, as of the late 1960s. This explains how, between 1968 and 2015, the peri-urban population increased from 9.4 million to 15.5 million and the peri-urban area increased from 19% to 25% of the national territory (Cusin et al. 2016). The growth of this population, which has been somewhat slower over the last 20 years, has come about both through the densification of peri-urban societies (i.e. by internal growth) and by absorbing an increasing share of rural societies (i.e. by external growth). In reality, this growth very often results from the integration of rural and urban territories under the direct influence of large communities. There are

51. Ibid. p. 126.

52. According to INSEE, peri-urban communities constitute the outskirts of large urban areas (with at least 10,000 jobs and where at least 40% of the active residents work in an urban center). It is therefore an extension of the city focused towards a less dense and more extensive habitat, which by its very nature devours space.

many reasons for this strong, sustained and sudden growth. Without going into detail, we will mention here the ones, which seem to us, to be the most decisive.

First of all, there is a problem related to the value of land and its evolution over time in large cities. Indeed, the scarcity of land in large urban centers, associated with an influx of population, leads to an increase in demand for apartments, the supply of which is stagnant. The adjustment will inevitably be made by the prices and therefore by the continuous increase in the square meter price of apartments, thus marginalizing a whole portion of the population who will be pushed back towards the belts that are far from the centers. In addition, the effects of speculation will accentuate this phenomenon.

During the "glorious 30 post-world war years", the explosion of the private car as an object of consumption and its systematic use, combined with the significant improvement of the road network and its adaptation to the exclusive use of the car, largely contributed to the phenomenon of suburbanization[53]. The fact is that all the measures taken to relieve congestion in city centers (construction of new roads, construction of bypasses, construction of parking lots in city centers, etc.) have in fact contributed to accelerating and accentuating the phenomenon of pressure exerted by the individual car: the more traffic lanes there are, the more cars there are.

Finally, suburban housing refers to life choices that can be expressed in two ways: on the one hand, it allows people to become private home owners; allowing people to own their own home and thus, in a way, control their future by securing a "roof over the head" as well as being a capital asset[54]. At the same time, it also allows for a "return to grassroots", that is, the possibility of choosing a way of life close to nature, with a garden; green spaces, less noise, in short, a quality of life that the larger complexes of the high-density suburbs were unable to provide to their inhabitants.

53. The evolution of the car fleet in France, up to the present day, is particularly significant, as indicated by the following figures: 1900: 1672; 1930: 1,400,000; 1950: 2,310,000; 1970: 13,710,000; 2000: 33,090,000; 2019: 39,910,000. With highest annual growth rates between 1955 and 1975.

Source: *L'industrie automobile française* – Analyse et statistiques 2019 – CCFR 2019.

54. With the sharp rise in unemployment resulting from the various economic crises that characterized the liberalization of the 1980s, the fact of owning one's home could constitute as a factor of security and a way of protecting oneself against the uncertainties of the future.

The question of the extension and continuous development of peri-urban housing could arise, in particular in view of the serious problems that this causes here and there: soil artificialization and the consequences that this can have on the environment, the continuous loss of UAA (Useful Agricultural Area)[55], the enormous uptake of private cars, fine particle pollution, problems related to water and its exploitation, etc. However, it seems that the substantial interest in the individual house is less the cause of this growth than the attachment of communities to certain large urban areas. Since the beginning of the 2010s, there has been a stagnation in the number of single-family homes and the number of primary residences as a whole. Finally, the construction of single-family homes is clearly slowing.

At this point, after having put into perspective the conditions for the growth of city in all its forms, it is necessary to appreciate the consequences, particularly in the relationship that it maintains with its mediate and immediate environment. In other words, can we consider that the emergence of cities whose population exceeds 10 to 20 million inhabitants (or even more in certain cases) does not lead to irreversible consequences that are likely to put the planet in danger in the long term? Is the current ambient discourse, mainly in the form of the threats from climate change, not the perfect illustration of an imminent danger for humankind and the planet they inhabit? We propose below to quickly address this problem in order to consider (or not) the possibilities of reacting to such an eventuality.

2.4.2. *From the Fordist city to the sustainable city: a gamble?*

The observation is clear: the emergence of gigantic cities with millions of inhabitants can only further manifest the problems related to pollution and environmental damage that could ultimately jeopardize the future of the planet. This is because the city alone integrates almost all human activities (with the exception of agricultural production), whose consequences on the environment are extremely significant. The idea of a sustainable city stems from the need to change this, thereby allowing for a city that would have (as much as possible) a healthy regard for nature, and life in general. Without going into detail, we will put these different problems into perspective.

55. It is considered that the useful agricultural area has been reduced by some 60,000 hectares every 10 years for several decades, which in the long term, cannot continue without posing serious problems.

2.4.2.1. *The Fordist city: an overview*

As we have seen, the intensive Fordist period of accumulation during the *Trente Glorieuses*, associated with the quantitatively important movement of populations, also had qualitative consequences whose extent we can measure today. In this respect, we can consider the city as the symbol of a mode of development, including its most negative aspects, which the dynamics of urban concentration in recent decades can only continue to exacerbate. A review of the pollution that the city causes, or rather organizes, leads to the observation that the urban environment, because it concentrates both a (very) dense population and multiple activities in a small area, is obliged to support many flows, namely: the distribution of water as well as its recovery, the circulation and flow of people and goods, the routing of energy, the removal and treatment of waste, in short, a colossal set of operations with all the real and/or potential risks vis-à-vis the health of populations. Without going into detail, we can summarize the threats of widespread pollution as follows.

2.4.2.1.1. Atmospheric pollution

The first form of pollution, from a quantitative and qualitative point of view, is atmospheric pollution resulting from the combustion of fossil fuels and which plays an essential role in the deterioration of the state of health of populaces, as well as the exacerbation of chronic pathologies, such as cardiovascular diseases, for example[56]. These are mainly activities related to transport (private, public, goods, by men and women, etc.), in all its forms, daily travel within the city as well as outside of it. Chronic exposure to air pollution can jeopardize the life expectancy of city dwellers due to its importance for public health: it is estimated that this exposure, for example, to fine particles, for 25 European cities (which represent a total populace of 39 million) is responsible for 19,000 premature deaths every year[57]. The unbearable image of cities such as Shanghai and Delhi, Mumbai and Beijing, which alone crowd together over a hundred million people and which are covered in a thick residue of fine particles, paints a terrifying vision for these cities' future inhabitants. In addition to local emissions of pollutants, there is also exogenous pollution, which is "imported" over very long distances and

56. For more on the subject, see the conclusions proposed by the WHO (World Health Organization): Burden of disease from environmental noise – Quantification of healthy life years lost in Europe, 2011. WHO, Review of evidence on health aspects of air pollution – REVIHAAP Project, 2013.

57. See: ANSES, *Radiofréquences et santé*. Mise à jour de l'expertise. Avis de l'ANSES, Rapport d'expertise collective, 2013.

whose concentration is particularly dangerous. Another consequence of transport activities in the city is the noise it produces, in combination with the noise of the inhabitants living there. Although noise does not incur the same concerns among city dwellers as air pollution, they rank noise concerns ahead of insecurity and pollution in the hierarchy of the most worrying problems of their neighborhood or their commune[58].

2.4.2.1.2. The water crisis

The so-called water crisis is not only due to a simple problem of availability, even if water scarcity can very quickly become a very real problem. If water, in principle, is an infinitely renewable resource, it is nevertheless finite in terms of its availability. This is mainly due to the fragility of the natural reproduction process, attributed to two main factors:

– on the one hand, as a consequence of appropriation conflicts by some to the detriment of other users (upstream/downstream relationship) and/or due to its unequal natural distribution on our planet;

– on the other hand, it is also susceptible to multiple pollutions, for which it is the object, and which are more often than not the costs of human activities.

The global volume of usable freshwater represents less than 0.01% of the water present on the surface of the globe, that is, 12,500 km^3 of water. This quantity would be globally sufficient for current uses (production and consumption) if it were equitably distributed, which is not the case, either in time or space. For instance, Europe has 8% of the available water for a population that represents 13% of the world population. These figures are, respectively, 15% and 8% for North and Central America, 11% and 13% for Africa and 31% and 60% for Asia.

However, the major problem at the origin of the water crisis lies in the unequal access to the resource: if the domestic water consumption is on average 40 liters per day per inhabitant, there are deep disparities: 10 liters per day for a Malagasy peasant, but 240 liters per day for a Parisian inhabitant who is not a peasant. In addition, 1.1 billion people do not have access to running water and 2.4 billion people do not have access to sanitation facilities. Under these conditions, 60% of childhood illnesses in the world result from infectious and parasitic diseases that are spread by

58. AFSSET, *Impact sanitaire du bruit. État des lieux*, Indicateurs bruit-santé, 2004.

water. One of the UN's goals is to halve the number of people without direct access to water and those without sanitation facilities by 2015. In order to achieve this, it would require the water connection of more than 370,000 people every day, and more than 820,000 people to sanitation per day. In other words: impossible. More generally, if we want to solve the water problem (all uses combined) by 2025, it would require an investment of 180 billion dollars per year… a gamble lost before it has begun.

2.4.2.1.3. Human activity

We could thus continue to unpack the extent of the damage done to the environment by human activity, particularly those occurring in cities. Whether it is the catastrophic deforestation in Brazil (and also elsewhere), destroying what has been called "the lung of the planet", the rise in sea level as a result of climate change, and its consequences on populations, the increase of natural disasters, the irreversible loss of biodiversity, we would not have finished summarizing the negative byproduct of human activity, and as such, this is not the object of our subject in this context.

However, at this point, it is appropriate to raise an aspect that to us seems to be decisive, namely, the importance of the social character of the measures, expected and hoped-for, to get the planet and the city out of what can be described as an impasse, should things continue as they stand. This is one of the strong points put forward by sustainable development (and therefore by the sustainable city): along with the economy and the environment, the social aspect of sustainable development is decisive. This is perhaps the stumbling block insofar as the social cost of the transformations necessary to respect nature may appear as a prohibitive cost to some, particularly in the context of a competitive "free" market economy.

Finally, all the phenomena we have mentioned above have increased with neoliberal accumulation, accentuating the gap between rich and poor and fashioning the neoliberal city that has been stamped with the seal of poverty and generalized insecurity. For emerging countries, poverty is crammed into squalid high-density areas, while the European city sees its entrances disguised as supermarkets and other places of exchange of goods and services. In any case, it is indeed the concern for the future of the planet that remains most important[59].

59. For more on this topic, see the remarkable work of Pinson (2020).

2.4.2.2. *The sustainable city: a solution?*

The question we asked earlier, "How can we live harmoniously together on this planet?" also means questioning the organization of human society, through a long-term lens, which takes charge of the answer right now, that is, by taking action. However, the mode of development that we know inexorably leads to pressure on cities, which have become disproportionate megacities in some cases. In 2004, there were 20 megacities, that is, multitudes of more than 10 million inhabitants, representing 4% of the world's population. In 2007, more than 50% of the world's population lived in a city with more than 500,000 inhabitants, including slums. Finally, in 2020, the 10 largest cities in the world will each have more than 24 million inhabitants.

The concept of the sustainable city is an attempt to provide a practical response to the challenge of overcrowded cities that serve as a refuge for the world's misery. Its design and construction are entirely rethought to reduce, as much as possible, the "ecological footprint", that is, the impact of its operations vis-à-vis the environment. It avoids urban sprawl and favors the economy of resources. Urban sprawl is particularly costly in terms of space consumption, as well as in terms of resources of all kinds, notably through the mandatory use of automobiles and their consequence on greenhouse gas emissions. This leads to a rethinking of city planning in order to implement a denser and better-quality urbanism: organization is rationalized to avoid any unnecessary travel and to facilitate soft traffic or public transport. This leads to a functional mix, as in BedZed (Beddington Zero Energy Development) near London. Each building is made up of homes facing south and offices facing north. This allows for two things: better thermal management and, for some residents, the opportunity to work onsite. The shops are also at the heart of the district and the social mix is conceived less by ideological choice than by functional necessity: it allows for the grouping of diverse skills and know-how and makes all the services necessary to everyday life available to the inhabitants. This strengthens the social bond between different social categories and weaves networks of solidarity. The social "benefits" of diversity appear to be inseparable from the functional benefits: it is not the juxtaposition of different populations in the same place that gives diversity its qualities, but the sharing and exchange of ideas that take place in the context of daily life.

Transforming a traditional city into a sustainable city is particularly complex, as we can well appreciate, and will certainly take a long time;

operating conditions cannot be changed overnight, as they obey a logic that has been developed over centuries and yielded to a logic of growth and development that took no account of environmental constraints. Under these conditions, only new businesses, newly renovated neighborhoods, new architecture, can directly integrate the desirable characteristics of a sustainable city. In this regard, BedZed-type neighborhoods are expected to multiply, and are already happening in Germany and China, as well as in Northern Europe and the United States. This does not prevent us from noting the relatively derisory nature of these attempts in relation to the global task.

However, a short-term action can already be envisaged and is proving relatively effective in terms of travel and construction. Indeed, as far as buildings are concerned, they are being designed in such a way as to greatly reduce the use of non-renewable energies, with a view to constructing "low-emission" buildings. However, the reduction in the flow of materials and resources does not only concern energy: water savings, for example, are more than 30% when using rainwater harvesting for sprinkler systems or sanitary purposes. In terms of waste recovery, the results are also very satisfactory, particularly with regard to a concept of waste production at the source, in the context of a circular economy and the questioning of a certain mode of consumption.

This being said, a major question arises: can what is done here and there, as highlighted above, be considered as generalizable at all times and in all places? Nothing is less certain, because if the technologies are mastered, the means to implement them worldwide as quickly as possible remains inadequate. In the meantime, palliative solutions must be found in order to postpone the deadlines. Urban resilience must be considered in this perspective.

2.4.3. *Cities and resilience*

As mentioned above, building the sustainable city will take a long time, given the magnitude of the tasks to be carried out and the cost of the operations. Making cities like Delhi, Mumbai and Sao Paulo "sustainable" cannot be achieved by a simple "snapping of the fingers". While waiting for "a better world", the concept of resilience as applied to the city appears to serve as a reference; it is in fact a matter of rethinking the concept of urbanity from a systemic point of view, when it is confronted with major disruptions that are likely to put the city in danger, as well as to recover all of its functionalities as a result of these same disruptions.

2.4.3.1. *Urban resilience: meaning and scope*

As we have seen above, the demography of cities continues to gallop along, so much so that by 2050 between 70% and 80% (depending on the source) of the world's population will be urbanized. Another equally important figure, according to the United Nations, indicates that the average cost resulting from natural disasters will be close to 250 billion dollars. Putting the economic cost to one side, we can expect the human cost to be far greater and more serious.

In this way, the concept of resilience as applied to the city results from this simple and not very engaging observation and has as its operational objective to respond to such a situation. Urban resilience makes it possible to think of complex urban systems according to the disruptions they may undergo, in order to overcome them and return to their initial state and once again function normally. Urban resilience therefore stems from the logic implied, or even imposed, by risk management. Urban resilience is part of a multidisciplinary construct, which itself requires architecture, of course, as well as energy management, sustainable construction, health planning, etc., that is, a systemic whole capable of withstanding various crises (climatic, ecological, physical, etc.) but that also protects the populations concerned.

This concept finds its primary meaning in the physics of materials and refers to the idea of the ability of a body or a material to resist a shock or deformation. However, very quickly, the concept became transversal and transdisciplinary, so that it took on a holistic aspect due to its multiple applications for biology, ecology, psychology, sociology, economics, etc. In addition, research into the urban environment, which has taken hold of the concept, also involves the use of a wide range of academic skills; the concept of resilience gives rise to numerous problems and methodological developments meaning that all the disciplines involved are necessary. Even if the concept appears in some cases to be redundant and lacking rigor, the objective remains to be more operational and functional than theoretical. Indeed, resilience, by definition, is considered as the possibility that the complex urban system contains, on the one hand, the ability to absorb a disruption but, on the other hand, able to recover its former functionalities as quickly as possible. As a result, we can distinguish two forms of resilience: on the one hand, a "short-term" resilience that must respond to the immediate dysfunctions resulting from disruptions; on the other hand, a "long-term" resilience whose objective is based on maintaining the main and essential functions of the city, that is, the quality of life, the functioning of heavy infrastructures, the attractiveness of the city, etc.[60] However, it is clear that to speak of resilience is to consider the complexity of the city in its entirety, that is, by taking into account all the elements, factors and variables that come into play in this same complexity, including those that have been invited into the debate, such as sustainable development. As a consequence of the multiple and widespread incidents against the environment, the concept of urban resilience makes it possible to put into perspective an approach to the urban environment in terms of its trajectory, the components and functioning of which can be based on the normative principles of sustainable development. The explosion in the number of natural disasters over the course of the last decade alone makes this particularly significant[61]. The best possible response to natural disasters that disrupt the city's functions is to address the causes and "origin" of these disasters. In this way, we can consider sustainable development to be a complementary response to the problem of resilience over time. In addition, resilience will play a significant role in the emergence of the sustainable city, that is, an "ideal" urban system in terms of its respect for the environment, which would significantly mitigate the seriousness of the disruptions we are experiencing. In terms of an extension of one by the other, resilience appears (or at least is presented as such) as a means for achieving the sustainable city. This is because the consequences caused by disturbance factors make it possible to put in place mechanisms that will not necessarily lead to the initial state but to an improved situation. The sustainable city as an expression of normative principles that likely results in a kind of "ideal city" would thus be the result of the resilience that makes it possible. Resilience as the capacity of urban systems to first adapt and then overcome various disruptions seems to be the best way to meet the operational needs of the sustainable city as an ideal city. It is therefore possible to consider that the "resilient city" and the "sustainable city" are inextricably linked in space. Resilience can thus be

60. We borrow from Toubin et al. (2012) this dual reading of urban resilience, which we consider significant from both a functional and operational point of view. See also Lajaunie et al. (2002) and Emelianoff (2007).

61. An overview of the site: https://www.catnat.net is revealing. For example, the site estimates that global warming costs Belgium nearly 10 billion euros per year. The data concerning the frequency and violence of storms, rising waters, bad weather, as well as drought is significant.

seen as an operational approach capable of meeting the challenges of sustainability as applied to the city.

As we can see, the resilient city/sustainable city articulation participates in and relates to an integrated management logic. Indeed, optimizing the resilience conditions for a city in a situation of major disruptions makes it possible, if the solutions are envisaged from the perspective of sustainability, to save time and space at the same time as responding to the challenges that will inevitably be caused by the ongoing global warming. That said, at this point, the question arises of assessing the scope as well as the limits of the concept of resilience when it is associated with the concept of integrated management.

2.4.3.2. *Limits to the concept of resilience*

The value of the concept of resilience as presented above is obvious, especially when coupled with the principles of sustainability. However, it is also appropriate to highlight its limitations and question its scope.

The concept of resilience can be criticized for opening the door to certain practices that can be described as passive, particularly in the search for sustainability. Indeed, the driver of resilience lies in the idea of a return to "normal", which in the context of disruptions is a return to the initial state. On the one hand, this is generally the initial state that is at the origin of these disruptions and their consequences. The case of global warming and the natural disasters associated with it easily demonstrate this. By prescribing a simple return to the former state, we allow for a situation that will continue ad infinitum. On the other hand, this observation causes some urban planners to refuse the idea of any evolution of the urban system by putting forward the central idea of resilience, namely, the desire to return to the previous state on the pretext that it is stable. In this regard, we could therefore consider that the slums that occupy a large portion of our megacities constitute an initial state whose existence should not be questioned. These remarks make it possible to highlight the fact that the operationality of resilience is only valid within the context of the sustainability principles applied, in particular, the reconsideration of this initial situation.

As an extension of the previous remark, the operational nature of resilience may lead to an attempt to limit the costs of protection by focusing efforts on the ability of urban systems to recover. The objective would then become, not to consider a way out from the top (i.e. by choosing sustainability and accepting the costs involved), but to recover from the disaster as quickly as possible without further questioning the deep-rooted reasons that have caused this state of affairs. This is a "cost/benefit" issue, and these approaches have led decision-makers to mitigate risks, particularly in the case of floods. The question posed concerns the articulation between levels of protection (and the escalation of protection systems) and the acceptability of risks, which are never zero: it is indeed the articulation between these two levels that is at the center of possible choices. If the referents of resilience are primarily technical, the moral and ethical arguments of sustainability are never far away.

Finally, as we have emphasized throughout this section, the concept of resilience refers to the functions of operationality, that is, the technical dimension to possible interventions in order to return to an initial state in an altered urban system. We are therefore in a situation in which we respond to a technical problem with a technical response. However, the problems raised by the question of urban system resilience go far beyond the technical response and raise questions on society as a whole. Perhaps we are touching on the limits of a concept without rejecting its meaning and scope.

2.5. The city under the regime of service accumulation

We have previously discussed the change in the regime of accumulation following the failure of the "Fordist mode" in favor of a model described as neoliberal. This transformation appears to be the result of the rapid and decisive development of information processing and the development of appropriate technologies, namely, NICTs (i.e. New Information and Communication Technologies). This evolution of the model cannot be considered as neutral, especially with regard to the impact it is likely to have on the city and on the city's way of being. In this framework, we retain two essential aspects that we feel are important to remember and which we will expand upon next:

– on the one hand, the development of urban and peri-urban peripheries, characterized by the decisive importance that the businesses developed there have taken on, most often transferring commercial activity from the center to the outskirts of cities, or even large towns;

– on the other hand, the impact by the development of information and communication technologies on the nature of the city itself, and even on the meaning that the city gives to it (or tries to give to it), which has been coined as the "smart city".

2.5.1. *From the center to the periphery or the peripheral expansion of commerce*

One thing is clear, most cities in developed countries (but not only in those countries) have taken on a very similar physiognomy in recent decades: the massive usage of the car, peripheral development of commercial activities, the development of a mode of occupation of the city mainly around large outlying commercial complexes, etc. This movement is the result, in addition to demographic causes, of the crisis brought on by Fordism and the abandonment of industrial activities in favor of developing countries with far lower labor costs alongside the immense development of commercial activities as a substitute for these productive activities and, finally, globalization through trade. Therefore, just as the ring road characterizes the Fordist city and the massive infiltration of the automobile into city centers, the emergence of a compact system of tertiary activities on the outskirts of cities has become the almost inevitable norm. This is also true for smaller cities and towns that have a relatively significant population density around them, also known as "employment areas". Two main aspects will focus our attention: on the one hand, the main characteristics of this peripheral development and the specific problems that are consubstantial with it, and on the other hand, the direct and/or indirect impacts on the conditions for the emergence of the sustainable city. What has rightly been called "commercial urbanism" in fact finds its origin and genesis in the neoliberal model of service accumulation, the content and main characteristics of which we have previously defined, particularly in the context of the urban model and its consequences for the sustainable city. Several remarks:

1) Beyond the purely aesthetic aspect, which is also negative, and which has often been called into question due to the ugliness of the suburbs that result from their anarchic development, we see that this form of urban planning is based on three main pillars, which constitute its skeleton, the development of "self-service", so-called "discount" commerce, and advertising that is most often considered aggressive. In addition, these areas were created from the extensions of the triumphant model of the "car is king", which imposed itself with Fordism, as well as due to the urban sprawl that is the consequence of the (relative) abandonment of urban centers, at least initially. If we add to this the effects brought about by size and

dimension as a driving force for commercial efficiency (due to "price effects"), we can better understand the success of these commercial zones. Moreover, the introduction of digital technology through "e-commerce" (in particular with the development of the "drive") has enabled and accentuated this trend. From the point of view of sustainable development, we quickly perceive the problems posed by these service installations, and this, in several respects:

a) first of all, some have pointed out this aggressively "ugly France", which characterizes almost all large cities: the inconsiderate deployment of billboards, architecture made of "boxes", the scattering of car parks for each and every superstore, etc.;

b) another negative aspect of urban sprawl, which is nowadays of major importance (and rightly so), is the artificialization of soil, necessary for its transformation into a parking lot and associated with the business it houses, in accordance with the "no parking, no business" principle;

c) in terms of employment, the widespread idea of strong job creation under these vast commercial plans may in fact prove to be false. Indeed, the commercial jobs created here are generally at the expense of existing jobs elsewhere, and the creation of jobs for some becomes a loss of jobs for others. In general, it is the jobs from small businesses in city centers that suffer the consequences. Therefore, as noted by Garcez and Mongin, at the end of a survey carried out in eight towns, "no project opened in the last three years has attained the levels expected, promised jobs are not necessarily transformed into actual jobs, however the jobs destroyed remain destroyed". This is because after having created more than 500,000 jobs between 1990 and 2010, commerce has since experienced a reduction in workforce;

d) one last point concerning the architectural quality of the buildings in these commercial zones, which are commonly called "boxes" and which, as we have already mentioned, are associated with parking lots. For if the box form is not in itself particularly objectionable, aside from the point of view of sustainability, the problem is quite different: in fact, they are generally closed hangar-style buildings that are particularly costly in terms of energy (heating, air-conditioning, internal and external lighting, parking lots, etc.) and their transformation for different uses would be economically prohibitive;

e) in addition, the different peripheral commercial zones give rise to strong disparities in profitability, depending on the intensity of the exchanges they host and the quality of these exchanges. Therefore, it is possible to find rental values ranging from €100 per m^2 to several tens of thousands of euros. As we are also witnessing a strong financialization amongst the major distribution groups, the profitability of investments varies according to the sectors and the intensity of competition between the different players. However, one thing is clear: operators in these sectors do not necessarily have an interest in investing in a building, nor architectural pitches likely to reduce profitability, which is already under severe strain.

2) We cannot here evade the question of reconverting these commercial areas when this happens (which we have every reason to think will be inevitable in many instances). With regard to the current urban context, which is trying, as best it can, to integrate the constraints of sustainable development, there are many who propose the conversion of the large surface areas dedicated to car parks to re-densify urban areas: limitations to urban sprawl, the end of converting UAA (Useful Agricultural Area) into "urban land" [62], that is, the end of an infinite expansion of the urban (and peri-urban) area.

Furthermore, and still in the context of converting these areas, if a site considered profitable operates under the required conditions, the price of land (i.e. the land rent that such a situation generates) will then act as a brake that will not allow for (or with much greater difficulty) the "liberation" of these spaces, especially if the objective is to transform them into social housing, and this for a simple reason: the financial return on the business (i.e. its economic value) is much higher than the financial return we can expect from social housing.

3) The problems raised above, in connection with these urban areas dedicated to commerce, may in the long term come up against additional difficulties that are more structural in nature. Indeed, since the beginning of the 2010s, on a macro-economic level, the purchasing power of French people and the turnover of the retail trade have typically evolved less quickly than the number of shops (and therefore the dedicated m^2); at the same time, the trade sector (in the sense of national accounts) is conversely taking an increasingly larger market share. So much so that in a global context where urban sprawl has reached its limits, where the consumption of agricultural

62. This state of affairs is all the more important since the phenomenon of peri-urbanization has become, in the meantime, a determining factor in the loss of UAA at the national level. In France, the UAA equivalent of two French departments is lost every 10 years. We now also know the consequences of such an evolution of urbanization reflected for in the context of "natural disasters".

land and the artificialization of soils are widely being called into question, we are witnessing the emergence of a new vision that calls for the reconversion of these complexes. Even if, from an economic point of view, the freeing up of land at a moderate price remains difficult (because the yield of commercial finance – and therefore its value – remains higher than that of housing), the current slowdown in economic activity and developments, in terms of consumption and distribution, suggest an involuntary closure in the medium term for certain businesses[63]. That said, we must be aware that such a scenario will only be made possible if, and only if, the new situation brings value and therefore wealth.

In reality, the question needing an answer no longer concerns commercial sector growth as much as a "decrease" in this activity accompanied by a treatment of commercial wasteland. The possible alternatives could be coordinated around a few essential points: on the one hand, reconsidering the land problem as a strategic variable in urban planning and not as a simple adjustment variable in the context of commercial real estate projects; on the other hand, developing a mode of mixed urban functions, both in peri-urban (or even rural) areas and in built-up areas, which are thus more densely populated. Finally, to reconsider the objective needs of the areas themselves, whether they are territories or areas for urban accommodation.

It can be seen, if we want to move beyond the present situation, that the work to be done is considerable and there is a risk of coming up against local interests that are likely to call into question all the measures necessary for its realization.

2.5.2. *Smart city and sustainable city*

As we saw earlier, service accumulation results from a transformation of economic activity, mainly from a functional point of view (and not from a structural point of view), by affirming the primacy of the new information and communication technologies, which thus become available to the tertiary sector for which they become the central element. Consequently, we can logically think that we will witness a transformation of the city which, with these new conditions, is adapting to the new forms and modalities of

63. Moreover, it would seem that the Covid-19 crisis could accelerate such a process, the trade sector, especially those qualified as "non-essential", especially small local businesses that do not have a sufficiently important financial standing. As a result, in the years to come, we risk witnessing a reorganization of the entire sector.

accumulation: just as Fordist accumulation led to a city constrained by the massive development of the use of the car, ICTs impose new forms and modalities of identification on the new city. For not only has the city become "smart", but the habitat itself is adorned with the ornaments of this same intelligence. We will highlight two essential aspects.

On the one hand, the particular meaning we give to the intelligent or smart city can be summarized as follows: the city's smart character corresponds to the elements of the city that give it a definite ability to meet the requirements for improving the quality of urban services, namely, economic optimization in the use of ICTs, a form of social equity in the use of ICTs (openness, adaptation and sharing, institutional relevance in the use of ICTs) and, finally, environmental mastery of all the components that call on the ICTs. As we can see, the city is becoming smart due to explicit references to sustainable development criteria.

On the other hand, the assimilation between smart city and sustainable city seems obvious insofar as the state of the planet does not allow us to dissociate the two qualifiers, which would be contradictory. The new definition of the smart city is thus: the quality of intelligence and the "sustainable" developments of the city correspond to the characteristics of the latter, thereby giving it the capacity to satisfy the system in terms of the requirements that can be summed up as follows:

– sustainability of the economic control of its development and the economic optimization of the use of ICTs;

– social equity in economic, social and cultural development in the use of ICTs;

– openness, sharing and adaptation in the social use of ICTs;

– environmental control of all the components likely to develop as well as all the elements that constitute the use of ICTs.

So, the economic, societal (social and cultural) and environmental requirements become the reference system to define and characterize the smart city. We can only note the sum of difficulties that appear insurmountable when we consider the means that would have to be implemented so that cities such as Sao Paulo, New Delhi, Bombay, Beijing and the many other megacities of the same nature can become smart cities in the sense defined above. An immense project! We propose to further analyze the considerable impact that this urban movement has had, not only on the immediate lives of the people concerned but also on the particularly negative significance it has for the evolution of the planet, in many areas, but especially from an ecological standpoint.

The problematic that we propose to develop below can be summarized as follows: the recent emergence of the concept of smart city stems from a double requirement:

– on the one hand, to introduce into "the lived social world" the technical developments which are likely to modify (or improve) the conditions of daily life, both in cities and in the countryside, in particular through the generalization of ICT use[64];

– on the other hand, to become a decisive factor of sustainability, ensuring the harmonious development of the planet and thus of the cities which compose it, which are generally the source for ongoing disruptions.

It is in this conflicting perspective between "technological progress" and the endangerment of the planet by human activity, and through the systematic deployment of these same technologies, that the concept of the smart city seems to impose itself, with technology at the service of humanity becoming the "*deus ex machina*" of a new world. The city appears as a highly technicalized artifact that should serve both as a basic substratum likely to ensure the "bright future of humanity" and, at the same time, as the main vector of the planet's sustainability that will protect generations to come. At this juncture, let us make a few remarks:

We hypothesize that the forms urban spaces take are the result of a particular historical periodization and that we cannot dissociate this form of urban composition from the socio-technical conditions of its production. Therefore, the forms of urban composition, the social forms of wealth production and the forms of accumulation of this wealth closely coincide. The city as an artifact can only exhibit the mode of production that produces and determines its content; the smart city does not depend on the characteristics of the technologies in use (even if they are ICTs), but rather on the intelligence of the mode of production that allows for it to emerge in certain forms and under certain historically defined conditions.

64. It is apposite to not systematize this kind of statement since, if this phenomenon did indeed take place, it did not exhaust the trade capacities of the city centers that for many have "resisted" what has all the same been considered a real threat.

1) In reality, if this city we are wishing for is smarter than the previous ones, would it just be due to the technicality that characterizes it and is expressed as a superior form of human mastery over nature? This would finally realize the wish formulated by Descartes, almost four centuries ago: "man as master and possessor of nature". This would therefore correspond to technology, the technical way of thinking, the ultimate reference for the characterization of "intelligence". Generally speaking, the question that arises here through the semantic effect of the "smart city" amounts to identifying a concept with a situation that consists in making us think of technology as the layout for our future, as the result of the intelligence of this and only of this. This amounts to forgetting that technology answers the question "how?", which does not deprive it of a certain form of intelligence, whereas science answers the question "why?".

2) Instead, we propose a systemic approach to the conditions of accumulation that brings to the fore changes in the conditions of accumulation that are leading us from an industrial society to a "service" society, with ICTs becoming, in such a configuration, the drivers of the change that must get us out of the rut. Therefore, the passage from a "Fordist" type of accumulation (1945–1975) to a post-Fordist type of accumulation (1980–2007) under the constraint of globalization and the financialization of the world economy, and then the post-crisis period of the "subprimes", all of which allows us to understand the reasons why "technical intelligence" is taking over from intelligence.

3) The apparent triumph of ICTs is not a surprise in itself insofar as it seems to be part of a humanist perspective that claims a form of moderation in the systemic logic of social production. This is particularly true of the perspectives that engineers and other service technicians proffer, notably to save the planet, with ICTs becoming the *Deus ex machina* of a bright future.

4) This process is based on the following observation: the progressive substitution of technology for thought contributes to eliminating the critical moments in the formation of the latter by sticking to a purely formal observation which implies, consequently, a loss of meaning concerning reality: with the smart city, we no longer answer the question "why?", which alone should animate us (i.e. in the strict sense of the word, give us a soul), but the question "how?". The digitization of the urban environment as a panacea and a promise of a bright future is therefore more of a problem than a solution.

2.5.2.1. *A semantic approach to the concept of "smart city"*

This expression actually refers to two signifiers the meaning of which is both familiar and confusing to us:

– on the one hand, the city whose commonly accepted concept is linked to the practice that we have of it even if its definition is common since it is accepted by all. Beyond the different forms that the city can take, a common character allows everyone to appropriate its meaning: the city is an artificial product which has nothing natural about it but which is the result of technologies in effect at the time of its development, and which therefore characterizes the different evolutionary stages of these same technologies.

– On the other hand, the concept of intelligence which we agree to recognize as a question of the quality of understanding, that is, of interpreting the signs which present themselves, of seizing with the mind the nature and meaning of things that explain and account for them. We distinguish intelligence from understanding because it appears more intuitive than discursive and because its practice is accompanied by rational elements of an associative or affective order: thus, if we grant a certain form of intelligence to the animal world, we deny it any form of understanding and/or reason.

So, what about this term the "Smart City"?

The first observation lies in the fact that we associate two logics, two different coherences that very quickly become contradictory for two reasons:

– On the one hand, to paraphrase Heidegger, the city is like the rose, in the sense "the rose does not know that it is a rose": in other words, the city does not know that it is a city for the simple reason that it does not think of itself; it does not have any understanding of self.

– On the other hand, the city as a technical result comes from the ability to understand what surrounds us. Not that technology is foreign to any form of intelligence, but this association amounts to bringing variables into concordance even though their respective meanings distance them from each other: as if the "wrench" were able to determine the way in which to screw and unscrew bolts!

This last confusion leads to another, namely, the confusion between science and technology. In this respect, current language tends to associate the two terms in expressions such as "scientific" and "technological" progress. However, it seems to us that there is a confusion of meaning insofar as science discovers while technology invents, for example, we discovered DNA, whereas we invented the wheel. Associating science and technology is certainly not an abuse of language, but it can lead to confusion as to the meaning of this association. Of course, if technology preceded science in the history of humanity, this in no way diminishes the importance that either has taken, with the other following in its wake. Hence, as much as the expression "bypass city" has a meaning and a homogeneity insofar as it constitutes a technical response to a technical question, the notion of a "smart city" applied to a historical situation, as opposed to other historical situations, becomes nonsensical: in what way is the city of the 21st century smarter than the city of Antiquity or the Middle Ages? The bypass city is indeed a technical response to the artifact of the singular city of the 20th century, and it is just as intelligent as the "trendy" city of our period. We think that this antinomy of reason arises from the confusion that is deliberately perpetuated between the scientific questions that the world asks itself and the answers it suggests, in the absence of wanting or being able to ask them in their entirety. The ideological disguise that the technicization of discourses allows, compounds the confusion between discourse (*logos* in Greek) and the technical approach to societal problems.

2.5.2.2. *Service accumulation and the smart city*

As we saw earlier, service accumulation results from a transformation of economic activity by affirming the primacy of new information and communication technologies, which are thus made available to a service society for which it is the central element. Therefore, it is logical to think that we should witness a transformation of the city, which is now adapting to the new conditions of accumulation. We will highlight two essential aspects:

– On the one hand, the particular meaning we give to the smart city can be summarized as follows: the city's smart character corresponds to the elements of the city that give it a certain ability to meet the requirements for improving the quality of urban services, namely, economic optimization through the use of ICTs, a form of social equity through the use of ICTs (openness, adaptation and sharing, institutional relevance through the use of ICTs) and, finally, environmental mastery of all the components that call on ICTs. As we can see, the city is becoming smart due to explicit references to sustainable development criteria.

– On the other hand, the assimilation between smart city and sustainable city seems obvious because, given the state of the planet, dissociating the two qualifiers would be contradictory. The new definition of the smart city thus means that the quality of intelligence and sustainable development of the city correspond to the characteristics of the latter, which in turn give it the capacity to satisfy the system of requirements which can be summarized as follows:

- sustainability of the economic control of its development and economic optimization through the use of ICTs;

- social equity in economic, social and cultural development through the use of ICTs;

- openness, sharing and adaptation through the social use of ICTs;

- environmental control of all the components likely to develop as well as of all the constituent elements through the use of ICTs.

Therefore, economic, societal (social and cultural) and environmental requirements become the reference requirement system to define and characterize the smart city. We can only note the number of difficulties that appear insurmountable when we consider the means that would have to be implemented for existing megacities to become smart cities in the sense defined above.

2.5.2.3. *From the eclipse of reason*[65] *to the triumph of technical thought*

The themes developed previously lead us to face a singular situation: the place occupied by "the technology" in the organization of our daily life is increasingly important, not only within the framework of daily occupations but especially in the perspective that becomes ours with regard to the way of thinking. The question seems to me to be extremely important, as the mathematization of forms has led progressively to the threat Horkeimer noted of the power of subjective reason in the form of what we call the technicization of knowledge (Horkeimer 1974). Horkeimer shows how the dichotomy that is created within Reason between Objective Reason (Kant would have said Pure Reason) and Subjective Reason (Kant would have said Practical Reason) ends up becoming, with liberal thought, an instrumentalized thought, finalized, in the service thereof!

---

65. Of course, we borrow from Horkeimer this figure of speech, which is not only a simple metaphor but also the apt description for an inexorable movement, the smart city being the most successful expression for the degradation of the Reason phenomenon. For further reading on this subject, see Horkeimer (1974).

> The more the technical knowledge develops and the more [...] man sees the horizon of thought and the more his activity shrinks, his autonomy as an individual, his capacity to resist the invasive technologies of mass manipulation, his faculty for imagination and independent judgment. The propagation of the Enlightenment through technical means is therefore accompanied by a process of dehumanization. The progress threatens to annihilate the very purpose towards which it tends in principle: the idea of the man.[66]

Indeed, if we observe the evolution of economic thought, the gradual transition from objective thought, from "political economy" through to "economic sciences", as displayed on the frontispieces of the triumphant University, is in fact only a dressing and a disguise: we have passed in the name of science from the question of "why" to the question of "how", that is, to an organization of thought that has become an instrument. We explain how young malleable minds have been converted to false science by using mathematics for one purpose only: to develop functional relations that allow us to maximize for whom a production functions, for whom a consumption functions, etc. As soon as the demonstration was based on mathematics, it was made to seem that it was based in science, well anchored in the formalism of an "efficient" thought, given that only the result counts. Therefore, the faculty could declare to the world that the "economic sciences" had finally been recognized, with even more force now that they had become technologies for the maximization of functions that were commonly distressing, even false[67]. Reason thus submits to objectives and it is necessary to ensure that these are achieved in the best conditions and at the lowest cost. Accordingly, "reason, having abandoned its autonomy, has become an instrument"[68]. We can thus see to what extent mathematical formalism, as it is emphasized in this case, and moreover with reference to the use that physicists have made of it, thus becomes the *deus ex machina* of an instance, whose sole purpose is to render illegible the reality of the social (societal?) problems to which, normally, science must respond.

66. Horkeimer (1974), p. 10 (author's translation).

67. On this subject, see the criticisms of Joan Robinson and the Cambridge School of macroeconomic production functions, in particular that of Cobb-Douglas. If there is fraud, it can be found there.

68. Horkeimer (1974) p. 30.

At the end of this presentation, it seems important not to make a conclusion so as to allow each and everyone to appreciate the new situation as it presents itself to us, humanity as the creator of our own destinies. The stakes have become such that the manipulation of the elements of reality that surround us end up overtaking us due to the feeling that controlling what we produce is impossible. Algorithms have existed since the dawn of time, but what we had not foreseen has come to pass: we have given ourselves the means of calculation, such that the productions of which they are capable exceed our own, and perhaps may be a threat one day.

2.6. A provisional conclusion

At the end of this journey, any conclusion that is too abrupt would not only be premature but insignificant, even misleading. If the city is an artifact, as we have said, if it embodies and shelters a technical system (in the sense of J. Ellul), the fact remains that each city has a soul, that is, it tells a story, and has a meaning that has been imprinted onto it over time. Therefore, technology, time and space constitute the vectors, the very supports that make it possible to better understand and analyze the city in the making and the associated challenges.

However, on the other hand, cities, all cities, face the same problems, the same difficulties and, in many respects, are similar. The history of Delhi is very far from that of Sao Paulo, and even less alike that of New York or London. Yet the problems that they face are all strangely similar, their respective contributions to the suffocation of the planet through the emission of greenhouse gases, for example, are identical, albeit dissimilar in terms of the degree: although the cities of developed countries have acquired a great advantage over time, this account carries a large debit in terms of their contribution to environmental damage. The city is a space for the occupation of nature by humans, and it is also the space in which they produce, reproduce themselves, in the time that they produce the city and "its embarrassments". If the economy dictates to the city the structure for the conditions that produce it, then the city in return comes to consume this product, thus restoring an organic unity that can be described as a productive system, the conditions for consumption resulting from the conditions of production.

We have seen what the rules of the economy impose on the city even though this city, in its inextinguishable desire to grow, itself imposes on both human and nature, that is, on the "naturalism of humans and to the humanism of nature". This dialectic that sees the erosion of one and the other, of one by the other, sends us back to limitations that should henceforth not be pushed back and not exceeded at the "risk of risking" humankind itself.

To conclude on the future of the city, that is, on this complex technical ensemble in the making, would not be within our domain, let alone our ambition. We have described the urban phenomenon here by framing (or attempting to frame) the meaning it carries and thus the future that will be ours and that of our children.

2.7. References

Aglietta, M. (1976). *Régulation et crise du capitalisme*. Calman-Lévy, Paris.

Aristotle (1990). *Leçons de physique*. Presses Pocket, Paris.

Arnould, P., Bouerandi, E., Gilette, C. (2009). Rural/Urbain. In *Traité sur la ville*, Stébé, J.-M. and Marchal, H. (eds). PUF, Paris.

Arrow, K.J. and Debreu, G. (1954). Existence of equilibium for a competition economy. *Econometrica*, 22(3), 265–290.

Aydalot, P. (1980). *Dynamique spatiale et développement inégal*. Éditions Economica, Paris.

Aydalot, P. (1985). *Économie régionale et urbaine*. Éditions Economica, Paris.

Baran, P.A. (1967). *Économie politique de la croissance*. François Maspéro, Paris.

Baran, P.A. and Sweezy, P.M. (1968). *Le capitalisme monopoliste*. François Maspéro, Paris.

Bourdieu, P. (1998). *Contre-feux. Propos pour servir à la résistance contre l'invasion néolibérale*. Liber, Paris.

Boyer, R. (1979). *La crise actuelle : une perspective historique*. CEPREMAP, Paris.

Boyer, R. (1986). *La théorie de la régulation : une analyse critique*. La Découverte, Paris.

Boyer, R. (2015). *Économie politique des capitalismes – théorie de la régulation et des crises*. La Découverte, Paris.

Boyer, R. and Saillard, Y. (2002). *Théorie de la régulation : l'état des savoirs*. La Découverte, Paris.

Braudel, F. (1979). *Civilisation matérielle et capitalisme*. Armand Colin, Paris.

Bricke, D. and Ibbitron, J. (2020). *Planète vide – Le choc de la décroissance démographique mondiale*. Les Arènes, Paris.

Charbonneau, B. (1990). *Le système et le chaos*. Éditions Economica, Paris.

Choay, F. (1965). *L'urbanisme, utopies et réalités*. Le Seuil, Paris.

Choay, F. (1996). *La règle et le modèle, sur la théorie de l'architecture et de l'urbanisme*. Le Seuil, Paris

Choay, F. (2006). *Pour une anthropologie de l'espace*. Le Seuil, Paris.

Cusin, F., Lefèbvre, H., Sigaud, F. (2016). La question périurbaine : enquête sur la croissance et la diversité des espaces périphériques. *Revue française de sociologie*, 57(4), 641–679.

Davezies, L. (2012). *La crise qui vient. La nouvelle fracture territoriale*. Le Seuil, Paris.

Davezies, L. (2015). *Le nouvel égoïsme territorial*. Le Seuil, Paris.

Descartes, R. (ed.) (1953). Les principes de la philosophie. In *Œuvres et Lettres*, Gallimard, Paris.

Di Meo, G. (2005). *L'espace social : lecture géographique des sociétés*. Armand Colin, Paris.

Donzelot, J. (2006). *Quand la ville se défait – quelle politique face à la crise des banlieues ?* Le Seuil, Paris.

Ellul, J. (1977). *Le système technicien*. Calmann-Lévy, Paris.

Ellul, J. (1990). *La technique ou l'enjeu du siècle*. Éditions Economica, Paris.

Emelianoff, C. (2007). La ville durable : l'hypothèse d'un tournant urbanistique en Europe. *L'Information Géographique*, 71(3), 48–65.

Gadrey, J. (1987). *Les enjeux de la société de service*. Presse de la Fondation Nationale des Sciences Politiques, Paris.

Gadrey, J. (1992). *L'Économie des services*. La Découverte, Paris.

Gadrey, J. (2003). *Socio-économie des services*. La Découverte, Paris.

Gidrac, C. and Giroud, M. (2014). *Villes contestées. Pour une géographie critique de l'urbain*. Les Prairies ordinaires, Paris.

Gorz, A. (1977). *Écologie et liberté*. Galilée, Paris.

Gorz, A. (2019). *Éloge du suffisant*. PUF, Paris.

Gottman, J. (1960). *L'Amérique*. Hachette, Paris.

Gottman, J. (1961). *Megapolis, the Urbanized Northeastern Seabord of the United States*. MIT Press, Cambridge.

Hahn, F.H. and Matthews, R.C.O. (1972). *Théorie de la croissance économique*. Éditions Economica, Paris.

Horkeimer, M. (1974). *Préface à Éclipse de la raison*. Payot, Paris.

Jacquart, C. (2005). Urbanisme et développement durable : y a t'il un urbanisme durable. In *Le développement durable : approches plurielles*, Veyret, Y. (ed.). Hatier, Paris.

Jammer, M. (1993). *Concepts of Space. The History of Theories of Space in Physics*. Dover Publications, New York.

Kant, I. (2021). *The Critique of Pure Reason*. Penguin Classics, London.

Lajaunie, R., Villalba, B., Zuindeau B. (2002). Le développement durable face au territoire : éléments pour une recherche puridisciplinaire. *Approches territoriales du Développement Durable*, Dossier 1.

Laudati, P. (2012). Nouvelles perceptions et nouveaux usages urbains par les technologies numériques : le point de vue de l'habitant. In *Villes hybrides et enjeux de l'aménagement des urbanités numériques*, Zreik, K. (ed.). Europia, Paris.

Le Diascorn, Y. (2006). *La mégalopole japonaise*. Ellipses, Paris.

Ledrut, R. (1976). *L'espace en question ou le nouveau monde urbain*. Anthropos, Paris.

Lefebvre, H. (1986). *La production de l'espace*. Anthropos, Paris.

Leibniz, A. (1994). *Système nouveau de la Nature et de la communication des substances*. Garnier-Flammarion, Paris.

Lipietz, A. (1977). *Le capital et son espace*. La Découverte/François Maspéro, Paris.

Lorenzi, J-H., Pastré, O., Tolédano, J. (1980). *La crise du 20ème siècle*. Éditions Economica, Paris.

Mumford, L. (1973). *Le mythe de la machine*. Fayard, Paris.

Mumford, L. (2011). *La cité à travers l'histoire*. Éditions Agone, Marseille.

Mumford, L. (2016). *Technique et civilisation*. Éditions Parenthèses, Marseille.

Newton, I. (2016). *Mathematical Principles of Natural Philosophy*. University of California Press, Berkeley.

Paelinck, J.H.P. and Sallez, A. (1983). *Espace et localisation*. Éditions Economica, Paris.

Paquot, T. (2015). *Désastres urbains – les villes meurent aussi*. La Découverte, Paris.

Pinson, G. (2009). *Gouverner la ville par projet, urbanisme et gouvernance des villes européennes*. Presses de Science Po, Paris.

Pinson, G. (2020). *La ville néo-libérale*. PUF, Paris.

Plato (2016). *Timaeus*, 2nd edition. Focus, Indianapolis.

Riboulet, P. (1998). *Onze leçons sur la composition urbaine*. Presses de l'École Nationale des Ponts et Chaussées, Paris.

Robinet, A. (1957). *Correspondance Leibniz-Clarke : présentée d'après les manuscrits originaux des bibliothèques de Hanovre et de Londres*. PUF, Paris.

Ségaud, M. (2009). Espaces. In *Traité sur la ville*, Stébé, J.-M. and Marchal, H. (eds). PUF, Paris.

du Tertre, C. (2009). Modèle industriel et modèle serviciel de performance. *Economies et sociétés*, 43(4), 643–662.

Torre, A. (1988). Relations industrielles et découpages spatiaux. In *Traité d'économie industrielle*, Aréna, R., Benzoni, L., De Bandt, J., Romani, P.M. (eds). Éditions Economica, Paris.

Toubin, M., Lhomme, S., Diab, Y., Serre, D., Laganier, R. (2012). Résilience urbaine : un nouveau concept opérationnel vecteur de durabilité urbaine. *Développement durable et territoires*, 3(1).

Veltz, P. (1994). *Des territoires pour apprendre et innover*. Éditions de l'aube, Paris.

Veltz, P. (2017). *La société hyper-industrielle*. Le Seuil, Paris.

Zreik, K. (2012). *Villes hybrides et enjeux de l'aménagement des urbanités numériques*. Europia, Paris.

3

A Territorial Approach to Urban Spaces between Sustainability and Liveability

Fabrice ESCAFFRE, Mélanie GAMBINO and Sinda HAOUÈS-JOUVE
LISST Laboratoire Interdisciplinaire Solidarités, Sociétés, Territoires, UT2J, CNRS UMR5193, Toulouse, France

Cities have never been as crowded as they are today, with half of the world's population living in urban areas. A large proportion of these areas, especially the largest ones, have seen their population grow over the last 50 years, while others have just maintained or even seen their population decrease. All of them, however, have been strongly transformed spatially.

Therefore, urbanized areas have developed from cities. These urban areas were developed around these, according to a radio-concentric model, as well as in support of the various transport networks in more archipelagic organizations. These spatial transformations of cities are largely linked to changes in the mobility of individuals and goods, which have continued to increase and diversify during this period. They are also largely explained by the globalized functioning of the economy (Sassen 1996), in which many cities play a pivotal role, but for which they are also poles of concentration and, in part, the dissemination of wealth.

By evolving in this way, urban areas have had a strong impact, both directly on their own scale and indirectly on the scale of the territories necessarily involved in their operations, their agricultural environments and,

Integrated Urban Environment Management and Resilience,
coordinated by Luc ADOLPHE. © ISTE Ltd 2022.

more generally, their natural environments. This can be seen from land artificialization, brought about by the development of infrastructure, facilities and housing. This is generally accompanied by negative effects on biodiversity. This impact can also be measured through the increase in atmospheric pollution linked to the concentration of activities, flows and housing that constitute cities. From a systemic perspective, it is in fact the environments of the entire planet that are affected. At the scale of urbanized regions, this perspective is embodied in territorial metabolisms (Barles 2017) with balances that are all the more fragile due to urbanization pressures.

The recent or ongoing transformations of the environments has a retroactive effect on urban areas and cities; from the multiple effects of climate change, we can take the situation of coastal cities faced with rising sea levels as an example. We can also mention the cumulative effects of global climate change and the phenomenon of localized urban heat islands (UHI). There is no shortage of illustrations to emphasize the extent to which the current transformation and functioning of urban areas call into question the environmental sustainability of their organization. This questioning, which has been going on for a long time now, has influenced urban policies in various ways, encountering resistance from dominant market logics everywhere. Nevertheless, within urban policies, as well as alongside them, new proposals for the organization of cities and, more broadly, of territories have emerged, developing concepts marked by a concern to limit their negative environmental impacts. Eco-districts, eco-villages and eco-hamlets are illustrations of these alternatives. They can also be embodied in different ways by urban agriculture practices or mobilizations in favor of "natural" urban commons, such as water, for example.

The habitability of urban spaces is also challenged by contemporary urbanization (Lefebvre 2009). This translates into immediate problems in addition to those posed from a sustainability perspective. Therefore, urban habitability becomes a problem when a large part of the urban population is forced to occupy unhealthy housing spaces that are located in dangerous areas (Rebotier 2016) or be without any form of housing at all. If cities are attractive as places of employment, this does not guarantee that they provide good housing conditions. The same is true when they appear to be safer or less dangerous compared to other places. These urban housing problems are not new, but their magnitude must be measured against the number of people who are now "urbanized". This scale must also be assessed in terms of the dimensions of urbanized spaces.

Habitability issues may also arise due to the remoteness or limited accessibility to parts of the population. Indeed, the peripheral growth of cities is not only the result of the desire of inhabitants to live farther away from the dense and polluted centers; it is also a consequence of exclusion, through housing price mechanisms, for those who cannot afford to find accommodation there[1]. In addition to the environmental impacts already highlighted, such situations also have social consequences. They keep parts of the population from accessing certain urban services, forcing them to spend a great deal of time commuting, the financial cost of which weighs heavily on their budget. These habitability issues translate very differently, both spatially and socially, in and around cities. While they weigh heavily on the poorest households, they affect middle-class households differently, depending on the country, city and era. In all cases, however, although obviously with differences, these result in the socio-spatial divisions of urban spaces that can frequently be considered as forms of de facto segregation (Oberti and Préteceille 2016).

Contemporary urban spaces cannot be completely reduced to the two categories of problems just mentioned. However, because the issues of sustainability and habitability are sufficiently significant, we have chosen to study these in greater detail, specifically looking at where they overlap. The questions posed by habitability in the short-term intersect with those that sustainability invites us to ask with regard to the long-term, all the while putting them in direct tension with one another.

We will successively analyze urban spaces from the point of view of the habitat and through the mobilities that animate them, before studying their relations with agriculture and interactions with the climate. Before doing so, however, we will attempt to specify the general outlines and specific dynamics of the urbanization process over the last 50 years.

3.1. Urban diversification and transformation

As a global process, urbanization refers to similar developments without totally erasing the specificities of each city or their territories. Among these specificities, and without being exhaustive, here are a few examples to

1. See: https://www.facebook.com/puem.mx/videos/2377613122253158 for the movie: "Leti et Elis: dos caras de un dia en el valle de Mexico" from the Pelikano collective.

highlight what the analysis of urban spaces gains by being attentive to local and regional scales (Magnaghi 2014).

3.1.1. *The criteria for necessary urban differentiation*

The first refers to the specificities of natural environments. It makes it possible to distinguish between cities according to their integration as part of a climatic whole, their topography, or their proximity to water. In a context where, as we have just emphasized, the issues of urban sustainability are more topical than ever, these dimensions are of vital importance. The situation of coastal cities faced with rising sea levels illustrates this as much as the situation faced by highland cities where natural risks (seismicity, effects from sloped inclines, etc.) are cumulative, increasing their vulnerability. The environment, potentially or actually, is also an asset for cities, who take advantage of these resources for their economy. Many tourist towns thus endeavor to make the most of their pleasant climate or base part of their activity on the export of agricultural products that depend on soil quality. Finally, both as a constraint and as an asset, the particularities of natural environments influence the spatial organization of cities, their development and urban planning. In this sense, they play a major role in giving an urban area its shape. For example, it is easy to distinguish between urban spaces developed on a flat plain, radiating almost unhindered 360° from the center, those located on a hillside that typically form of a semicircle, or those organized linearly along a mountain valley. Therefore, from these three characteristic situations, Toulouse in France is different from Algiers in Algeria, and both differ greatly from Medellin in Colombia.

The second specificity that can be highlighted stems in part from the previous one, and also refers to the specific urban history of each city. It relates to the urban context in which a city is embedded. Depending on the country or region, cities are part of urban networks that are more or less dense, hierarchical and centralized (Burgel 2003). They thus function with varying degrees of autonomy and complementarity in relation to other cities, whether nearby or further away. Thus, some capital cities are at the head of an urban network, for example: Paris. Such networks can be very stable over time. They are directly linked to the organization of political power, but can also be analyzed on the basis of economic relations between cities. Political and economic dynamisms do not necessarily overlap, given that political capitals are not always the most economically influential cities: Washington versus New York or Los Angeles in the United States, or Quito versus Guayaquil in Ecuador, for example. The development of urbanization, however generalized it may be, does not put an end to such specificities, but rather articulates them and finds points of support there. If it generates transformations, it only rarely totally calls them into question.

Thirdly, urban spaces are also distinguished on the basis of their economic profiles and trajectories (Veltz 2014). Between a city where administrative functions dominate, compared to an industrial city, commercial city or tourist city, to name but a few types, the landscapes vary, as does the population. Each type of city also undergoes different developments according to the prevailing economic dynamics. For example, many industrial cities in the northern hemisphere have experienced difficulties in recent decades as deindustrialization affects the economy of their country or region, becoming increasingly oriented towards the tertiary sector, services, technological innovation, etc. That said, many urban areas, particularly the largest ones, have hybrid profiles. These must therefore manage the diversity of their economic orientations. This can be a guarantee of attractiveness and even help stability. It also presents the need to organize activities that have very different characteristics together.

Cities and urban spaces are also highly differentiable in terms of their populations and settlements. The social, economic, cultural and political profiles of the inhabitants vary from city to city, but are also often diversified and unevenly distributed within them. For example, university towns have on average younger populations, whereas there tends to be more older people in seaside towns, especially in the off-season. In urban areas, suburban or peri-urban sectors may be strongly marked by the presence of the middle classes, as well as young couples with children on occasion. Finally, certain neighborhoods, and even certain cities, bring together culturally related populations, as in the case of quarters inhabited by Asian populations, frequently referred to as "Chinatown" in the largest cities. This analysis of the social diversity of cities is inexhaustible and refers not only to the categories of inhabitants that it highlights, but also to their more or less strongly differentiated lifestyles. These can, for example, clearly influence the commercial apparatus or be inscribed in its religious monumentality. These can also result in de facto inequalities, more rarely reflected in laws. They are strongly marked, for example, by the housing sector, and also result in unequal mobility access, as previously mentioned. They can also take many forms, such as segregated schooling, inequalities in access to health care or exposure to natural hazards, etc.

Finally, a last element of the specification of cities refers to modalities in their planning and management. Urban plans have been widely developed, but they do not necessarily exist everywhere, or are differently precise or restrictive. Urban governance (Le Gales 1995) is hence weighted differently depending on whether the weight of public actors and market actors is stronger or weaker (Harvey 2010). It also depends on the place that it leaves for civil society, or the place that civil society chooses to occupy. Administrative and political divisions also vary from country to country, in relation to the different definitions of cities and the urban phenomenon. Moreover, culturally, the relationship with private, public or collective ownership of land can differ greatly from one country to the next, and this has a major impact on urban policies. When we add all of these aspects, linked to urban public action, we understand that they specify cities, at the very least at the scale of large global regions, but frequently also within these areas, depending on the country or region.

The vectors differentiating cities and urban areas are therefore numerous and potentially very powerful. Beyond the global elements analyzed in this chapter, they call for in-depth methodological reflection. This is limited, on the one hand, by the necessary questioning of the comparability of urban spaces, and, on the other, by the limits of strictly monographic analyses or analyses limited to a particular type of city. We will not return to these dimensions in this chapter, but make mention of them as a prospect for further study, complementary to the analyses that make up this chapter.

3.1.2. *Urbanization as a global phenomenon*

With variations, urbanization is developing as a global phenomenon. It began with the Industrial Revolution and has continued ever since, especially after World War II, at different rates and to different extents in different countries. This urban growth is closely linked to concomitant transformations in the countryside, which has lost some of its inhabitants, and is sometimes repopulated as soon as the process of urban sprawl sets itself in motion (Kayser 1997). In all cases, urbanization corresponds to the extension of the urban domain in a relatively short time. It rarely results in the appearance of new towns, but rather to the growth of existing cities, within the perimeter through densification, as well as beyond, through the extension of urbanized areas.

At the same time, it takes a more diffused form since it also engenders a certain standardization of lifestyles that can be seen in particular through consumerism, as well as through relationships to the media, and today, through digital technology, and even as a result of the daily commutes and residential mobility that connects territories more than ever before, making them function as a network (Bourdin 2019). This last aspect invites us to emphasize how much this increased urbanization is linked to globalization, particularly in terms of the economy. The flow of tourists between cities illustrates this, first and foremost the spaces that welcome them, such as airports. This can also be observed, after much more research, through the flows of energy and matter that transit and feed the urban metabolisms that frequently extend over very long distances to supply cities with food, energy, etc.

Over the course of the last half-century or so, urbanization has been particularly active around the largest cities, to the point that a set of concepts around the metropolis, notably metropolization (of the region or metropolitan area), is sometimes used to analyze it (Ghorra-Gobin 2015). This conceptual field refers, at first glance, to the economic and often political strengthening of the world's largest cities. More specifically, it deals with the networking of territories and urban sprawl from the main cities. It is used to analyze the situations of global cities, as well as those of secondary cities, also on a world scale and which are nevertheless large cities in their own right. With metropolises, we are in fact dealing with one of the key points of support for the global economic system. This can give rise to pro-metropolis analyses and actions, but has also generated research and critical stances towards metropolises (Faburel 2018). Independent of this, the economic, demographic, political and, indeed, significant environmental impacts of metropolises characterize contemporary urbanization. They can even, when their multiple relations with other territories are taken into account, serve as a matrix for the analysis of several other dimensions of urbanization.

This is true of urban sprawl. Indeed, metropolises are often only presented as territories that polarize people, activities, infrastructures and wealth, forgetting that they are also characterized by significant growth on the outskirts. Whether this growth is due to the shortfall of jobs, the desires and/or constraints that push inhabitants to live outside city centers, or the result of urbanization from the main transport axes and nodes, the horizontal growth of metropolises takes place on the outskirts, resulting in the urbanization of areas that were still agricultural or natural. Already, built-up sectors undergo selective transformations according to their forms, positions, activities or pre-existing settlements (Charmes and Keil 2015). In all cases, many metropolises build or reinforce the existence of urban regions. Within them, cities and countryside are no longer in stark opposition. A third space, an “in-between city” (Sieverts 2004), acts as a buffer between them, modulating landscapes and activities, in a multitude of hybridizations and differentiated articulations.

Urban sprawl or peripheral development is not, however, the prerogative of metropolises. It is in fact observed around cities of various sizes and influence. It reflects the major impact resulting from the increase in individual mobility capacities and the effects of urban segregation, as previously mentioned. The phenomenon is widely criticized for its contributions to greenhouse gas (GHG) emissions, often too accelerated (Newman et al. 1995; Dupuy 2002), as well as the land artificialization that it generates. It also supports an important part of the housing economy and, more broadly, the residential economy (Davezies 2008), while expressing aspirations for a life that is closer to nature than that permitted by inner cities, especially within larger ones. It is therefore no easier to find a consensus on its definition than it is to define metropolization.

We previously emphasized that urban contexts can differ greatly from one region of the world to the next, and from one country to another, but their common characteristic can be seen in their stability over time. The processes we have just analyzed do not therefore modify them totally, but they do affect them. This is the case, for example, with the strengthening of large regional cities in Europe (Le Gales 2011), which have benefited as much as they have suffered from metropolization, especially if we consider that they are still largely under the influence of the few European global cities. Moreover, among these intermediate cities, those with a mono-sectoral and industrial profile have had a difficult time so far, especially those that developed more recently, in particular the socialist countries (Burgel 2003). The stratum of small- and medium-sized cities is also concerned by the effects of these dynamics. Here again, some of them benefit, such as those that enjoy both touristic appeal while being near enough to a metropolis. Others, on the other hand, suffer when they have an industrial profile, are isolated and, moreover, experience residential competition from territories located on their immediate periphery. In many countries, the parts of the urban framework that are in crisis have become major political issues, above and beyond issues surrounding land use planning and urban development.

Behind the urbanization of the world in recent decades, there are not only territorial differences, but also a variety of processes that depend on the dynamics driving them or the territories in which they take place. The analysis of urban spaces according to a clear and stable typology, particularly a hierarchical one, increasingly seems to have to integrate the interactions between territories, both near and far, beyond inherited territorial categories such as town and country. From this point of view, questioning the habitability and sustainability of such urban spaces requires constant reasoning on several scales, from the local to the global, as we will now do by studying the five interfaces mentioned above, beginning with an analysis of the one which is organized around housing.

3.2. The urban issue through the prism of housing

Housing and habitat issues provide an angle for the analysis of urban spaces, showing how much they are torn between sustainability and habitability at the first stage. They refer to a strategic sector in various ways. Of course, because it corresponds to a primary human need, it is often considered primarily in its economic and legal dimensions. These dimensions intersect, in particular, around the notion of private property, making housing a market good of primary importance, both in terms of household budgets and at the macro-economic level, as well as, albeit with nuances depending on the country, as a heritage asset. In addition to this diversity of meaning, there are those that make housing an object of public policy, particularly urban policy, as well as those that invite us to observe it through its multiple relationships with architectural and urban forms. The urban evolutions already mentioned can be reread through the prism of housing issues in the strict sense, but can benefit from a more extensive life-focused approach.

3.2.1. *Urban housing in transformation*

This notion places the inhabitant at the center of the analysis of urban spaces (Lefebvre 2009), on the domestic scale – at the residential space, the neighborhood (Authier et al. 2007) or the village (Charmes 2019) – by placing it within a system of place practices (Stock 2004) and thus directly linked to questions of mobility. The concept of inhabitation is then extended to include the uses, perceptions and representations of space in their diversity and globality. Living is understood as constructed on a daily basis, or more exceptionally, through a perspective that does not completely separate housing from other urban functions, but seeks, on the contrary, to integrate them, understand their connections and go beyond this by taking an interest in the lived, sensory, social, represented and even imaginary dimensions.

Over the last 70 years, the relationship to housing and urban living patterns has been greatly transformed. Housing patterns have evolved with large housing estates, suburban housing, the renewal or decline of heritage buildings and, more recently, new neighborhoods undergoing "renewal", or expansion sectors, close to central areas or connected to public transportation, etc. Sometimes, however, there are also neighborhoods or even entire sections of urban space composed of informal, self-constructed and more or less precarious or consolidated housing. Beyond that, modes of living have also been animated by residential movements. These movements, both move-ins and move-outs, whether forced or chosen, and all the possible gradations this implies, invite us to re-examine the anchoring of inhabitants that we generally associate with housing. In recent decades, the settlement of urban spaces has been nourished by new anchors, more or less permanent, and others renewed by the rural exodus or the urbanization of the countryside close to cities.

In addition to contextual explanations, these general developments also have a lot to do with the entry of the economy into a post-industrial and globalized phase. Some populations and territories have thus been weakened residentially by the socio-economic crisis that has accompanied this change, while others, who have benefited more from this evolution, are nevertheless also partly affected by it, with access to housing becoming increasingly difficult. Cities and their surrounding urban areas are thus faced with differentiated levels of residential attractiveness, ranging from the very low in shrinking cities, to the very high in those where the economic situation and general living conditions seem to be better. Urban areas undergoing strong growth must therefore meet significant housing requirements, often in short periods of time and with varying levels of land use regulation. Such situations tend to improve habitability in general, but with very strong social inequalities. Consequently, the most modest inhabitants are often forced to live in the most precarious and least comfortable housing, in sectors that are devalued, whether these are central or peripheral. Among the constraints that weigh on their living conditions, the progressive awareness as to the effects of urbanization on the environment makes it possible to underscore the extent to which the most disadvantaged live, in urban environments that are less frequently of good quality, and more often than not, at the other end of the spectrum, subject to natural and industrial risks or pollution (Larrère 2017).

Beyond these contextual factors, further than individuals and social groups, questions pertaining to housing and habitat must also take into account general social developments. Take, for example, the pursuit of individualization (Elias 2008), social networking and the weakening of traditional "integrating" institutions, which are some of the elements impacting this change, with important nuances depending on the region of the world. This affects the residential sphere, resulting in a reduction in the size of households and, consequently, an increase in housing requirements. In addition to this process, which is particularly visible in the large cities in countries of the North, there is an unprecedented demographic situation marked by the aging of the population, as well as by the migration of people to these countries and cities. Similarly, urban modes of living are not immune to the more general process of social insecurity that characterizes the contemporary period (Beck 2001; Castel 2003). A strong need for reassurance through the territory is just as perceptible as numerous forms of fragility (Bourdin 2015). Among these, settlement inequalities place inhabitants in contexts of unequal access to housing, services, jobs or a healthy living environment. In addition, this generates, within the social categories that have the capacity to do so, forms of residential withdrawal that can be observed in all parts of the world, with the process of residential enclosure according to the model of the gated community.

3.2.2. *Housing policies in question*

Faced with such issues of habitability, public action is acutely scrutinized. This is the case locally, at the level of cities and urban regions, which can enact agendas through their general planning choices or more specifically, in the field of housing. It is also at the national or even supra-national level, where laws or measures are developed to regulate or simply support the development of the commercial housing sector. With connections, as well as possible contradictions between these levels, public action in the field of housing can seek to eradicate inequalities or, through a mitigation perspective, aim to make them acceptable (Donzelot 2012). We may wonder about the weight of public action in the face of the financialization of urban growth. Which actions does it employ to intervene; between a sectoral or more territorial approach; between management and urban projects? Finally, at what scale should it be considered, when, as we pointed out at the beginning of this chapter, it seems almost illusory to precisely delimit urban spaces and the economy when modes of living give pride of place to network logics?

Without giving precise answers to these questions, since there are so many urban variations, we will simply point out that housing policies, where they exist, have undergone phases of change in recent decades that are inextricably linked to contemporary urbanization. Starting in the 1960s, these range from their support for the development of large housing estates, to their ambivalent relationship with suburban housing, from their concern for the improvement of heritage housing, to their intervention in self-build sectors, to their recourse to the massive production of private housing. Depending on the country and the city, they also refer to different conceptions of social housing. While it may refer to rental housing, as in France, for example, it may correspond to support for home ownership, as has been the case in recent years in Brazil and Mexico, for example. It is also possible to include all of the various forms of participatory or cooperative housing, as long as they are outside the real estate market. At the interface between urban planning and social action, the actions of public authorities in the housing sector should also be considered in their multiple economic implications. This sector weighs heavily on the banking system. It also affects both the macro-economic and the local construction economy. The subprime crisis and the bursting of the real estate bubble in Spain in the mid-2000s are reminders of the importance of these links.

Housing and habitat constitute a complex system of actors in which public action plays a more or less central and active role. To stick to the theme of sustainability and habitability of urban spaces at the heart of this chapter, it should be noted that this is questioned everywhere, in terms of the support it provides to urban sprawl and its forms or to the maintenance of existing housing. It is also challenged everywhere by issues pertaining to the housing of the poorest populations. We can also observe how much it integrates, for example, and the growing concerns related to the energy consumption of households. Often, in relation to the previous perspectives, it sometimes opts, as with the eco-districts in Europe, for a perspective that considers housing globally in relation to environmental issues. Exposure to risk, health issues or access to nature linked to housing are then taken into account. The preceding points outline the contours of a housing policy that we can imagine is implemented with some degree of difficulty. Its confrontation with short-term market rationales is one of them. Another is its confrontation with social inequalities, of which the residential sphere is one of the main expressions, as well as an important vector for its reproduction. A third is an inertia in the modes of living. Sometimes explained by very structuring cultural dimensions of societies, today this inertia can also result from real estate and urban marketing, which also influences residential choices.

The challenges of housing and habitat policies are specific to urban areas where residential attractiveness is high. This attractiveness accompanies the demographic growth of urban areas where it is being observed. It even makes it particularly consistent and rapid. It is thus characterized by increased competition for access to good housing conditions. This is reflected in situations where the average price of housing, both for rent and purchase, is generally higher than elsewhere. This is also reflected in the low proportion of unoccupied, vacant housing.

Overall, this tends to improve access to housing for the part of the population with the least resources, but in the case of the most attractive areas, it also puts the middle classes in difficulty. Accordingly, if we observe in these territories of high residential attractiveness many new or renovated dwellings that are generally expensive and sought after, for both housing and tourism, we likewise observe many situations of poor housing, ranging from shantytowns to over-occupied dwellings, otherwise taking the form of very long social housing waiting lists. These situations mainly affect the most vulnerable, but they also sometimes affect the middle classes. In addition to less favorable housing conditions, it is more difficult for them to access property. Sometimes this leads them to move away, sometimes far away from the city so that they too can become homeowners, at the cost of high travel costs and commute times.

In such contexts, housing policies have to deal not only with the issue of access to housing for all, but also with the effects of strong residential attractiveness according to the distribution of social categories. Indeed, in addition to the remoteness of certain populations, in a context of high tension, we observe opposing phenomena of residential upgrading or downgrading. On the one hand, the most highly valued sectors tend to increase in value and expand their urban environment into neighborhoods or territories that were previously less valuable. This corresponds to the phenomenon of gentrification (Chabrol et al. 2016). At the other end of the spectrum, some spaces suffer from long-lasting relegation, such as certain large social housing neighborhoods or vast self-built areas. The latter sometimes remain devalued because of their peripheral location, location in a high-risk area or lack of public transport services. Strong residential attractiveness increases the effects of exclusion and implies that housing policies should not only fight for access to housing, but also for social diversity, which can sometimes be difficult or even contradictory. How can the housing conditions of the poorest populations be improved without increasing the price of their housing or the value of their territory, which would ultimately exclude them?

The answers to this last question are all the more complex as housing policies have to deal with the urgency of rapid demographic growth, while taking into account the issues of sustainability applied to housing and urban space. When this is attempted with weak public land control, the constraints on urban public action are maximal. Therefore, for example, the response of rapid and massive housing production, as observed in a number of developing countries, is not free of perverse effects ranging from the standardization of the housing produced to its location and sometimes its lack of public transport access. These productions, however necessary they may appear, are not always accompanied by the maintenance of neighborhood relations, which we know are fundamental and go beyond the consideration of housing as a simple building and an economic good. In a different way, in certain rapidly growing cities in developed countries, the massive production of housing calls into question the urban fragility that accompanies it. Here, we find the same questions relating to location and service, but we also question the capacity of these cities to not only be attractive spaces, but also places in which the inhabitants are permanently involved, anchored and invested. In addition, as per the two previous examples, housing policies are also challenged to produce all of these new homes quickly, while minimizing the recourse to urban sprawl, thereby limiting the environmental impacts of cities implied. The inertia of such urban growth trajectories is great and therefore constitutes a major challenge for both the sustainability and habitability of urban spaces.

In urban areas where residential attractiveness is low, housing policies are confronted with issues that are partly different from the previous ones. Indeed, the population generally does not increase, or if so, only slightly, and sometimes even decreases. Therefore, there are many unoccupied dwellings and this generally helps to keep housing prices relatively low, at least compared with those in cities with high residential attractiveness. These two dimensions may suggest that there is no housing problem in these cities. This is not the case.

Therefore, while on average access to housing is easier for everyone, their lower prices are related to the often-lower incomes of the populations living there. The low residential attractiveness can be explained, for example, by the lack of job opportunities in a city or the limitation to low-paying jobs. Similarly, not all housing conditions are good for everyone. The low value of housing may be due to the inability of the inhabitants to maintain it properly. This is particularly the case when they are owners but their income remains modest. This low valuation can also be explained by the low income that renters can obtain from their property, which does not encourage or allow them to maintain it sufficiently. Therefore, in cities that are not very attractive from a residential point of view, there are neighborhoods that are particularly affected by the concentration of these low-quality housing units, as is the case in many of the downtowns of shrinking cities in France. Other parts of these cities, on the other hand, offer good housing conditions, particularly for populations belonging to at least the middle classes. The cheaper housing and land also allow inhabitants to live in larger areas in well-maintained homes. However, this often comes at the price of the need to use a car, as the low residential attractiveness is also frequently linked to poor public transport services.

Because of this, as well as because of the obsolescence of some of the housing, these urban spaces of low residential attractiveness are also subject to specific sustainability issues. How can we enhance the value of housing that is the least attractive and already built, located generally near the city centers, in other words, those that present advantages with the perspective of limiting land consumption and land artificialization, and may also be economical in terms of construction materials? And how, in doing so, can we generate enough value without excluding the low-income populations living there? How can we respect the heritage while carrying out work to improve the habitability of housing, as well as its energy performance and comfort? These few questions thus make it possible to emphasize that housing policies do not only concern the most attractive cities, but that they must be considered on the scale of the diversity of urban situations specific to each country or region. In these contexts, these policies cannot be implemented on the scale of housing alone. In general, they call the habitability and sustainability of urban spaces as a whole into question.

3.3. Urban spaces and mobilities

This can be seen by considering that contemporary urban development is inextricably linked with the growth of all types of mobility, from local to global scales, for people, goods and information. Focusing here mainly on the mobility of individuals, it should be noted that it takes place on several temporal scales. That of residential pathways, in other words, the changes in an individual's housing during their life. Secondly, the scale of their everyday life, in which movements connect the different places around an individual's home. This obviously includes places of employment, consumption, education, leisure, etc. Finally, the complex temporalities of leisure and tourism also generate travel. In the case of leisure, they are embedded in mobility and daily life. In the case of tourism, these are more exceptional and often longer in terms of their spatial scope. Cities and, more broadly, urban spaces are marked by these different forms of mobility, which play a large part in their recent transformations and reflect the issues of sustainability as well as habitability in other ways.

3.3.1. *Residential mobility*

Contemporary residential mobility is bringing about major transformations in urban structures, city spaces and, more broadly, in territories as a whole. Therefore, the residential mobility chosen by the most affluent part of the population largely explains the phenomena of heliotropism and coastalization that have been observed in recent years in various countries. They are studied in particular on the basis of changes in the residence of retired people, but they have a wider influence on all social categories. Other inhabitants are affected by their residential mobility. We can think of those that emigrate from deindustrialized cities in the space of a few decades or those from situations of armed conflict or violence. Whatever the cause, these mobilities, or even residential migrations, recompose the territories of the cities they leave, as well as those of the places they settle. In both cases, they are questionable in terms of sustainability. Consequently, in the cities that have left, to what extent does the departure of the population accompany or even accentuate a more general urban decline? To what extent is it economically, socially and environmentally sustainable? Can public action limit this, and if so, how? From another perspective, does this decline not open up new opportunities by de-intensifying cities, relaxing their real estate markets, and allowing more room for nature? For the cities and regions that have been favored in recent years, the analysis of the positive effects of their attractiveness is not sufficient, even if it is generally privileged. Indeed, when this attractiveness is strong, it generates rapid growth that feeds part of the economy, but it also has its darker sides. Access to housing, but more broadly to facilities and services for all, is not always able to keep pace with growth. This growth can be driven by the lure of profit or by the need to minimize its environmental impact. Therefore, urban development also strongly challenges urban planning in its willingness and ability to regulate.

Beyond these strictly urban aspects, the residential flows that animate cities also lead to transformations in local societies. In some cities, the departure of young people increases the percentage of elderly people. Elsewhere, the arrival of rural populations partly transforms the urban social organization. Elsewhere again, it is a question of cultural or religious diversification. Cities, as host societies, are challenged in their hospitality beyond their mere attractiveness. Individuals and groups who have recently arrived in the city are challenged in their methods of urban integration. The question of residential mobility thus goes beyond that of planning and is deepened by integrating social and political dimensions, as well as anthropological dimensions on the modalities and conditions of the city-making process (Agier 2015) in order to develop society.

Within residential flows, how are cities today places of passage or residential anchorages? How does this vary on the scale of urbanized spaces that have become vast and diversified? Some parts of these spaces are more specialized in the temporary reception of populations, while others are more specifically specialized in the settlement of populations. But which areas are involved? Reception is partly facilitated through the presence of rental housing, but it can also correspond to the presence of available land, for example, for self-construction. In other words, it can take place in the center, as well as at the margins. The interface between urban spaces and residential mobility is therefore not easy to read. It is nonetheless central in that it allows for an analysis of the movement of spaces in time periods that are admittedly relatively slow, but potentially very transformative because they are linked to a multitude of individual decisions.

Therefore, many of the environmental impacts from urban spaces are relevant to such decisions and movements, and we cannot consider analyzing them or even trying to reduce them without trying to act upon them. However, for each individual, the forces behind these movements are extremely powerful and therefore very difficult to regulate. Therein lies a main tension in the urban space–mobility interface.

3.3.2. *Daily and leisure urban mobility*

The second, and more familiar, tension relates to daily mobility. These are indeed essential to the functioning of cities and urban spaces. They connect them with their environments at different scales. They animate them with internal flows that link their different places and centralities. Daily mobility, if we remain focused here on the mobility of people, responds to various motives. We often observe urban functioning in terms of homework flows, but we also know that these are not necessarily the majority of the many other reasons for moving around the city. These different reasons may give rise to different organizations of mobility, depending on whether they are organized in a chain of flows or consecutively from a particular point, usually housing.

The variety of urban mobility can also be understood in terms of the modes of travel they use. Preferred modes of travel vary from city to city, but the automobile is consistently present and generally dominant. Alongside this, urban dwellers make more or less use of so-called soft or active modes such as walking or cycling. They also use public transport in all its diversity, from buses and coaches to streetcars, subways, trains and even cable cars. These modes, which are the main ones, are available in many variations and are complemented by more specific offers or uses, such as self-service bicycles, hoverboards, and scooters that have recently appeared. All these modes are used either independently of each other or in an articulated manner. Therefore, multimodality is a central aspect to be taken into account in the analysis of urban travel. This is becoming all the more prevalent given how ICTs now make it possible to optimize the management of connections between modes.

The diversity of motives and modes of transport are combined and made visible in cities and urban spaces by the infrastructures, networks and places that people use to get around. Some are dedicated spaces by destination, such as subway tunnels, or by adaptation, such as dedicated bus lanes. Other spaces are shared by several modes of transport, as well as by different uses, such as public spaces, especially streets and squares. Mobility also marks the urban space from the centralities that are often the hubs or nodes of transport systems. We are thinking here, of course, of railway stations, airports and main public transport stations.

Mobility and its spaces are not, however, found in all cities or in all parts of urban spaces in the same forms. Therefore, large cities generally have the best connectivity to a variety of transport systems. This is one of the key attributes of metropolises. However, given their size and the diversity of spaces that make them up, they reflect very different situations. Their main center is generally the place where the greatest diversity and density of mobility is observed. On their peripheries, for example, walking may be very low. Some of their secondary centralities may mainly operate from a metro station or from public transport more generally. In smaller cities, the modal mix is less diverse, particularly in terms of transport requiring heavy infrastructure. Of course, as previously pointed out, these differences between cities or parts of urban areas are compounded by differences in the income of the inhabitants, as well as their different relationships to mobility. Indeed, mobility is not considered by all in the same way. It calls for representations, as well as skills that are specific to individuals and partly linked to their original social environment. Therefore, daily travel, like changing our home during our lifetime, is an act whose social dynamics must not be erased in favor of a strictly spatial analysis, or one that only takes the supply of transportation systems into account.

Without claiming to be exhaustive, let us analyze in more detail three dimensions that are brought into tension by the interface between urban spaces and everyday mobility. First, the relationship between mobility and urban forms. The increase in daily mobility capacities has largely contributed to the de-densification of cities, with their horizontal development complementing, but sometimes replacing their vertical growth. The low density that was long antithetical to the city has thus become one of the urban components. The result has been an increase in the distances traveled each day by the inhabitants to connect their various places of life. In this context, the spatial organization of urban spaces has also been modified. From a model mainly structured around a main or even unique and necessarily multifunctional centrality, urban spaces have become polycentric. This has accompanied a triple evolution. Designed and developed according to a principle of functional specialization, space has seen the emergence of multiple centralities, clustered and connected along transport axes. The latter are primarily automobile-based, but the structuring principles described above also operate in more modally diversified transport systems. Such organizations have allowed the urbanization of new spaces, often accompanied by their commercial development. In some urban contexts, such as medium-sized cities in Europe, this has contributed to the weakening of other spaces with a higher density, as was the case when cities were mainly experienced at a pedestrian scale. This "fragmented" city is now experiencing contrasting developments on a formal level. We can observe the continuation of horizontal growth dynamics, certain forms of densification and functional diversification, as well as transformations in the organization of transport and mobility. This diversification can take place, for example, as a result of the development of public transport infrastructure whose stations become densification poles. It also corresponds to the ever-growing influence of digital technology on the organization of daily life, both in terms of optimizing mobility and remote work options. It is also sometimes based on the rediscovery of active mode advantages, such as bicycles. These observable changes, which vary from one city to another, are not taking place without some measure of difficulty, given that urban areas have been developed over the past few decades according to the logic of urban sprawl that was largely based on automobile mobility, de-densification and functional specialization.

Such rationales, which are still dominant, lead us to emphasize the effects of mobility on the urban environment in the broad sense. Three aspects, among others, can be mentioned. Firstly, mobility requiring infrastructure contributes to the artificialization of land in and around cities. This can impact natural or agricultural land. Landscapes are modified. Certain risks are increased, in particular those linked to water runoff, which can lead to catastrophic flooding. Mobility is also one of the sectors that contributes to CO_2 emissions into the atmosphere. This is largely due to the proportion of motorized vehicles using petroleum-based fuels. But it also refers to the daily distances they cover, which are increasing due to low density. However, on this last point, the debate becomes even more complex if we include not only daily mobility, but also occasional travel (Dupuy 2002). In addition, automobile mobility generates various types of air pollution, particularly fine particles, which have an impact on the health of inhabitants. Given the interweaving of contemporary urban dynamics and mobility, these negative environmental effects are major challenges for urban policies. They oscillate between maintaining the dominant mobility model, adjusting it or amending it more radically. These oscillations relate to the prioritization of the effects, particularly the economic effects, that are feared as a result of drastic transformations. These are also due to the now better-known health consequences of urban air pollution. These are also increasingly explained by a global environmental awareness. However, there are many contrasts between cities and countries. Some cities, such as Copenhagen, are leading the way in terms of cycling, while others, such as Paris, have only recently begun to move in this direction. Contextual effects are also extremely important, as shown by the cities that are trying to redirect part of their mobility to public transport after decades of being largely dominated by the car. Some cities, such as Medellin in Colombia, have even made this one of the flagship projects of their urban marketing, since the achievements, in this case a public transport network connecting the metro and urban cable cars, are so important and transformative. Despite this type of infrastructure, the decades of self-dependent urbanization are far from being erased. They call for long-term action to promote all alternatives to the car and coordinate their implementation with other urban policies.

Finally, and the achievements of Medellin make it possible to illustrate and question this, mobility also influences social relations in the city. Mobility calls for a first level (re)thinking on public spaces and their uses, since a certain number of them are subject to mobilization specialization (Escaffre 2020). By giving pride of place to the automobile, many public spaces have seen the flow take precedence over the station or exchange that they are supposed to allow. How many squares or streets have been partially or completely sterilized by automobile domination? Mobility then influences social relations from another point of view. They make it possible to link distinct places that are sometimes very distant as a result of urban sprawl. In this sense, they raise the question of access to the city and its amenities for the greatest number. In urban spaces with socially differentiated populations, this accessibility, or lack thereof, is one of the components of the right to the city (Lefebvre 2009).

This role of public spaces as meeting places and this question of urban accessibility therefore intensifies the other urban and environmental impacts of mobility. How can we avoid, for example, relocating low-income populations located in polluted but central parts of urban areas to areas that are certainly less polluted, but at the same time more distant or less well connected? Symmetrically, how can we avoid increasing the costs of automobile mobility beyond what is bearable for the inhabitants of the urban periphery, some of whom are modest or even poor, without giving up on reducing the role of the automobile? Such issues can also be seen when we consider the effects of increasing the value of real estate or even overvaluing portions of urban areas that have been reorganized to make less room for the automobile.

The last aspect of the complex relationship between cities and mobility that we will address relates to the place occupied by leisure-related travel in general and tourism in particular. Indeed, until now, we have mainly considered mobility in terms of its relationship with the general dynamics of urban spaces and, more specifically, with the uses of the latter by their inhabitants. However, one of the most notable contemporary urban phenomena refers to the place occupied in cities by leisure mobilities, and even to the fact that leisure and tourism explain certain forms of contemporary urbanization. Large passenger flows justify large long-distance transport infrastructures that have recently often been the object of sustained attention by urban authorities and private investors. These particular forms of mobility also make it possible to understand the choices made in favor of pedestrianization in many city centers. They certainly reflect a concern to reduce the use of cars, but often in order to increase the urban comfort of tourists, as much as the residential comfort of residents (Bourdin 2019). Finally, tourism has also led, through its development in the mountains or on the coast, to the generation of particular forms of urbanization, often monofunctional, accompanied by the transport infrastructures that link them to the rest of the territory. Accordingly, in addition to the environmental impacts incurred by this type of development, there is growing uncertainty as to their economic sustainability in the context of climate change. In different ways, the development of tourism in such territories is a major factor in urban transformations. In some cities, these transformations are so extensive that they have been challenged in recent years by social movements, as in Barcelona, for example.

The issues of housing and mobility, which have just been discussed, are closely linked and remind us how much the analysis of urban spaces, from the point of view of sustainability, benefits from being anchored as closely as possible to the practices of the inhabitants. In particular, such an analysis makes it possible to highlight the extent to which the challenges of sustainability include, but are sometimes also in contention with, those linked to a socially and territorially very differentiated or even unequal habitability.

3.4. An urban agriculture...

From the interface between agriculture and urban spaces, our analysis will expand to looking beyond the practices of inhabitants to a sector where contemporary urban transformations interact in multiple ways. If it is a key sector economically, it is above all essential from the point of view of the subsistence and health of individuals. Directly linked to land use, its spatial components, foremost among which are land ownership and landscapes, interact with urban spaces that are newly deployed in the surrounding countryside.

Before World War II, agriculture as a source of food underwent rapid modernization and specialization to become productive and thus allow for the deployment of a new agri-food industrial sector. This transformative movement in the countryside also had repercussions in cities and urban areas. The disappearance of a certain type of agriculture, in the market garden belts, for example, and its related activities such as slaughterhouses, commercial halls, etc., was such that by the end of the 20th century, cities were no longer considered as food production spaces (Salomon Cavin and Niwa 2011).

Today, agriculture and cities can be brought together again and re-articulated, so much so that many French-speaking authors (Banos and Sabatier 2010; Soulard et al. 2011; Le Caro et al. 2017) observe a hybridization of the urban and agricultural rationales that urban agriculture implies (whether in terms of actors, partnerships, methods of action or regulatory frameworks). English authors speak of integrated approaches, particularly in the context of territorial planning (Sonnino 2009; Mansfield and Mendes 2013). Indeed, since the 2000s, many cities have implemented food policies, which has led to a greater interest in the agricultural products that are present or to be promoted within their borders. The signing of a Food Pact in Milan in 2015[2] by more than a hundred cities shows that there are "proactive reconnections" (Poulot 2016) taking place between cities and their agricultures.

This favorable context explains why agriculture is no longer confined to production basins. Due to urbanization, it is now inserted, on the one hand, into various territorial systems, from cities to the countryside and, on the other hand, is present at all scales of living, from the building to the urban system, from the home to the territory, passing through all the intermediate scales of the street, neighborhood, suburbs and surrounding countryside. This context is also one of health crises (1990), the debate on GMOs (2000), the "hunger riots" (2008), as well as a search for sustainability in agricultural practices for the necessary implementation of the ecological transition. Demographic projections for 2050, which predict a world population of nine billion inhabitants, 80% of whom will be located in urban areas, have also contributed to the creation of a "new food equation", in which cities and

2. Today, there are 206 city signatories.

agriculture develop reciprocal functional links. The juxtaposition between cities and agriculture is finally being questioned by an agricultural world facing numerous challenges. In addition to adapting to climate change and the necessary ecological transition of agricultural models, there is a decrease in agricultural land and the number of farmers, the aging of farms, the low level of agricultural income and the need to diversify the more fragile sectors. The search for sustainability in agriculture is now accompanied by a search for both spatial and social proximity between farmers and non-farmers, one of the expressions of which is the re-articulation of agriculture with urban spaces.

Many local authorities in urban and peri-urban areas (from the "regional to local" levels) are also reinvesting in the specific issue of food. They are forging links with players in the agri-food market (dissemination of public purchasing, local supply in collective catering), and promoting local agriculture in order to develop new, more transversal food strategies (TFS[3]). However, these also rely on agriculture to slow down urban sprawl, ensure continuity that is essential for biodiversity and/or meet the strong increase in demand for recreational, educational or landscape services from their inhabitants. This growing reconnection between agriculture and cities makes urban agriculture a promise, capable of redesigning the city by proposing an attractive imagining, to reconcile city people and nature through food. The question is how urban agriculture can make the city more sustainable and habitable.

3.4.1. *Multifunctional urban agriculture*

The notion of urban agriculture can be understood in two ways. The first is to see it as all agricultural production activities carried out in the heart of cities, in the center of urban areas. The second, which is favored here, perceives it as all agricultural production activities present in the peri-urban areas of cities and metropolises (Charvet and Lureau 2018). It is an agriculture

> located in the city and its periphery, whose products are destined for the city and for which there is an alternative between agricultural and urban non-agricultural use of resources; the alternative opens up competition, but also to complementarities between these uses: built land and

3. A territorialized food system (TFS) can be defined as a "set of agri-food sectors meeting the criteria of sustainable development, located in a geographical space of regional dimensions and coordinated by territorial governance" (Rouillé d'Orfeuil, Rastoin 2015).

> agricultural land; water for urban needs and irrigation water; non-agricultural work and agricultural work; household and industrial waste and agricultural inputs; coexistence in the city of a multiplicity of skills due to migrations, cohabitation of agricultural and urban activities generating negative (theft, nuisance) and positive (green spaces) externalities (Moustier and Mbaye (1999), author's translation).

This definition presents three main characteristics of urban agriculture: location, consumer market and alternative uses of resources. These are forms of production that make the urban agricultural approach part of a territorial food system (TFS).

Other dimensions of urban agriculture have been identified by Duchemin et al. (2010). The spheres of intervention spontaneously identified are food security addressing food self-sufficiency (quantity and quality of food), the environment, including the protection of urban biodiversity, and the flow of matter and energy within the urban ecosystem (introducing/maintaining heat islands, provision for composting, water recycling). Others are education (raising awareness, personal and political empowerment), urban planning through citizen appropriation of vacant or green spaces, management of contaminated soils, economic development and poverty alleviation (aiming for economic inclusion), recreation by providing space for relaxation and contact with nature, health by allowing for outdoor physical activity and food quality, and finally… social interactions.

These multiple spheres of intervention directly or indirectly influence different aspects of the quality of the living environment for the inhabitants. Urban agriculture is a set of production activities that does not only provide food products, but which is also fundamentally multifunctional. It therefore links agricultural and urban planning issues. Thus defined, urban agriculture is the place where the articulation of cities with agricultural activity is manifested, as well as a process that recomposes both cities and agriculture. Several concepts have been used in scientific work to describe this: the long-standing concept of agri-urbanism (Vidal and Fleury 2009), the agri-urban project as a new alliance between the city and the agricultural sector, which promotes the agricolization of the city visible (Poulot 2014), and fulfils the agrarization of the city (Ernwein and Salomon Cavin 2014). The commonality between these notions is to underline that the links between cities and agriculture that facilitate the implementation of new agricultural, food and urbanistic practices.

3.4.2. ... *But a diversified agriculture*

Allotments, market gardening belts or small cultivated plots in self-built neighborhoods are old and now classic illustrations of the close link between agriculture and cities, which is also reflected, although at a different scale, in the garden city model. Floating farms, rooftop or facade cultivations, gardens, eco-pasture in the city, peri-urban farms, indoor and high-tech systems are all the forms that can be observed today. The factors explaining and/or characterizing this diversification of urban agriculture and the craze that accompanies it are numerous.

First of all, the location of the farm, such as its intra- or peri-urban location, the precariousness or sustainability of the land use, the size or type of space used (fields, wastelands, gardens, roofs, basements, walls, etc.) determines the techniques adopted and the organization of the actors involved in a given agricultural project in the city. The proximity and links with the city accentuate this diversification because they allow for the production of a wide range of products, the organization of production and marketing methods and also the provision of multiple services. Secondly, the type of products includes non-food products (ornamental plants, biomass for energy, services rendered) and food products (of plant or animal origin). From a technical point of view, the forms range from open-air and open-ground production to production under highly controlled artificial conditions with many possible variations and corresponding investments. From an economic point of view, the forms can be commercial or not. Not all will have economic profitability objectives.

In addition, production methods and marketing channels are becoming more diverse. A mix of models is often found within the same company. The most direct possible sale from producer to consumer is often sought. This can be done in a short circuit with a single intermediary or in a longer distance circuit. However, local circuits can involve several intermediaries within a given territory. Open-air markets, farm pick-up, farm stores that sell on behalf of a group of producers, AMAP[4] and basket deliveries are among the most common forms of marketing products and are not necessarily mutually exclusive.

4. An AMAP is an association of consumers tied to a local producer by an agreement providing for the bulk delivery of food products grown by organic agriculture on a set date, that is, community-supported agriculture.

Finally, the actors involved and the insertion in the socio-economic network also determine the forms that agriculture can take in the city. The degree of professionalism varies from leisure activities to complete professionalization, with intermediate stages. The networks of actors involved thus vary greatly and may involve farmers, local authorities for land management or economic development, urban planning agencies, chambers of agriculture and commerce, resident or environmental protection associations, researchers and, of course, project leaders… This last factor indicates that a common denominator could be that all these forms of agriculture are agricultures that are no longer organized around the single figure of the farmer.

3.4.3. *Six types of urban agriculture activities*

In view of this diversity, it may seem difficult to precisely identify the initiatives and combinations that take shape when agriculture and the city come together. Taking into account the space used, the integration with the built environment and the technology used, it is possible to distinguish six types of activity.

The first type is that of peri-urban farms, which are rather traditional, but which sell their produce through short circuits. Their difficulties are linked to the spatial position of these urban farms, often located in open spaces threatened by other uses. The urban rent may seem more profitable to build and store products, even when the agricultural land is valued by SIQO[5] (Perrin 2009). For peri-urban farms, the absence or weakness of land governance around cities and metropolises poses a problem. It is not guaranteed in the long term, while the investments made to set up new productions and new techniques commit to them for several years and can only be amortized over fairly long periods. This is even more pronounced in metropolises or urban regions where competition is not only over agricultural land, but also over access to water. Other technical problems characterizing this type of urban agriculture is the difficulty in circulating agricultural machinery, especially where the flows are dense and numerous, and where the plots are surrounded by axes of traffic circulation.

A second type includes so-called high-tech urban farms using technologies that aim to control growing conditions (urban greenhouses, the

5. Logos recognizing the origin and quality of agriculture and food products.

floating farm in Rotterdam, vertical agriculture promoted by Despommiers or in the basements of buildings in containers). This type of agriculture is experiencing problems related to the recruitment, in urban areas, of a workforce with the necessary levels of qualification adapted to the needs of this type of agriculture that is based on highly technical methods. There is also competition over wages for unskilled labor, which can find higher wages in the city in many other activities. Specifically illustrating a topic previously discussed, the problem of housing for this labor force is also significant, as rent can be higher in the city than in rural areas. In addition, projects of this type involve significant economic and environmental uncertainties: energy consumption, economic profitability, environmental standards, de-seasonality and deterritorialization. However, these limitations are not enough to invalidate this model. The optimism of investors and the significant fundraising by some startups as of 2015 could even herald their entry into a period of real industrialization.

Then, there are the so-called low-tech urban micro-farms producing in small spaces or on rooftops without the intensive use of technology, which are much less publicized, but well accepted by the local populations and actors of the territories concerned. This is the case for the upcoming largest cultivated area on a roof in Europe, located on the fourth floor of the *Parc des Expositions* ("Paris Exhibition Center") in Paris, which allows residents to have access to a plot of land for 320 euros per year.

These three types share a common difficulty, namely the organization and supervision of producers (IAU 2015). To solve this problem, some municipalities have favored agricultural projects that allow for the creation of "agricultural parks" with specialized agricultural sectors that integrate production and marketing (Madrid, Baix Llobregat in Barcelona, Ruhr, Milan), or agricultural projects located on high ground or on the walls of buildings, which are broadcast in the streets via giant screens (Shanghai), in order to publicize production and bring producers and consumers together. These are thus forms that integrate leisure activities to offer spaces for relaxation and social interaction.

A fourth type concerns cultivation in closed environments, corresponding to shared, family, individual or collective gardens. These are not a recent phenomenon, but testify to the long-standing (and sometimes forgotten) links between cities and agriculture. The experiences of these productive urban gardens do not all respond to the same dynamic. They can range from self-consumption to simple leisure to a real commercial activity, through a project that aims to restore social links or open up spaces for contact with nature in the broadest sense.

A fifth type refers to gardens in micro-spaces or the greening of urban spaces, such as balconies and sidewalks with the famous greening permits. The "Incredible Edibles" initiative in England is a good example of this type, since it invites everyone to use public spaces to sow and grow vegetables, which are then made available to everyone free of charge.

Finally, there is urban agriculture for leisure purposes or landscape maintenance with forms of eco-grazing. Despite recent experiments in eco-pasturing (e.g. the urban shepherds of Foix), the difficulties of maintaining livestock in the immediate vicinity of cities indicate that the range of agricultural production available to cities remains limited.

As these last three types demonstrate, cities are building an image of "fertile cities", green and inclusive metropolises, pushed in this direction by inhabitants who consume foodstuffs, but who also prioritize the preservation of green, open spaces and minimum built-up landscapes. However, agricultural actors do not necessarily wish to be associated with these amateur practices, with these non-market models, as they wish to defend a profession that they already consider endangered. Consequently, they may be suspicious of so-called urban agriculture projects that would justify, for example, under the guise of developing urban vegetable gardens, downgrading agricultural areas to building zones. Their fear is that these types of urban agriculture will diminish concerns that have more to do with the economy and/or agricultural policies, as can be seen in Europe around the Common Agricultural Policy and its recent developments.

The diversification of urban agriculture forms has resulted in a change in the way agricultural spaces are viewed from the city and by development actors. They are now mobilized as tools for regulating urban sprawl, as well as a support for the supply of green spaces. Nowadays, "agri-urbanism" has a certain number of specific tools at its disposal, such as the agri-urban programs in France (Poulot 2011) and PAENs[6]. Other conditions are necessary for the sustainability of these types of urban agriculture. Land and building regulations are also a key factor. Public awareness and

6. PAEN or Périmètres de Protection et de mise en valeur des Espaces Agricoles et Naturels périurbains: perimeters for the protection and enhancement of peri-urban agricultural and natural areas.

involvement, as well as the acceptance by the inhabitants of different systems and the products resulting from these systems, are decisive. From an economic point of view, the income and recognition earned by the different distribution channels or the organization of the profession are important.

3.4.4. *The establishment of food governance*

New actors in agriculture, often from outside the agricultural community, are increasingly involved in urban agriculture. This has greatly impacted the interdependencies and dialogue between agricultural and urban actors. Reactions to recent statements by the Paris City Council on the establishment of an agricultural cooperative to allow Paris to directly produce its own food illustrate the harshness of this dialogue and the difficulty of bringing about agricultural governance in the city. It is not self-evident due to the fact that the action of cities on the organization of their food supply has up until now been indirect and because agricultural actors do not want to be considered as simple contributors to the food and living environment of city dwellers. If these reciprocal and functional relationships become too unbalanced, the promises of urban agriculture would probably not be fulfilled. The dialogue between agriculture and the city can both enrich and impoverish the narrative potential and collective imagination that urban agriculture inspires.

In addition, food production is still not very present in the tools or procedures of territorial planning and urban development This gap forces local authorities and agricultural project leaders to work together and negotiate an emerging territorial food governance that can be beneficial to all urban agriculture stakeholders. In America, this is called urban food planning. This concept integrates agricultural and urban planning at the scale of an urban territory. It includes protected agricultural areas, particularly in the context of new green belts (Ottawa, Montreal), agricultural parks (Ruhr, Milan, Barcelona) or agri-urban programs (Paris, Montpellier, Lyon). In France, this is also reflected institutionally in the PAT[7]. They invite local authorities to take up food issues and invest directly in food relocation, by orienting, for example, local agricultural trajectories (promotion of short circuits, aid for the construction of direct sales outlets or on-site, requalification of wasteland, or even the creation of market garden plots in Lille).

7. Projets Alimentaires Territoriaux: Territorial Food Projects.

3.4.5. *The sustainability challenges for urban agriculture*

Ultimately, whatever their nature, the types of urban agriculture are almost all actions and projects that refer to the sustainable development paradigm. However, there are many questions regarding the sustainability of this agriculture. Some technological forms materially and above all symbolically deny the link to the land, the soil and the *terroir*[8], and yet are very popular with investors and politicians alike. The acceptance of urban agricultural forms by the inhabitants, however, points towards models that include social or environmental functions, as opposed to models that are only productive or technologically intensive. The discourse on high-tech urban agriculture, on farms favoring soilless techniques, carries ambiguity (Bonin 2020): if cities can become agricultural, why worry about preserving the land? Moreover, "soilless" farming does not offer landscape amenities, nor the present or potential ecological functions of agricultural land. Depending on the agricultural model chosen, the ecological and sustainable dimension is not the same, and can contribute to the illusion of a sustainable city.

In France, the ADEME[9] puts forward a few indicators to measure the sustainability of urban agriculture approaches. These are:

– the use of environmentally friendly agricultural practices that are part of agroecological or organic farming practices;

– the inclusion of the project in the circular economy and in the urban metabolism, whether it be the recycling of organic waste effluents from composting or methanization;

– the integration of the project in the territorial food projects (TFP) and in the supply of local products to the population;

– the development of links with consumers through educational activities and the development of direct sales;

– the participation in eco-designed networks to develop eco-design, for example, the optimization of packaging and transport;

8. Terroir is a French term used to describe the environmental factors that affect a crop's phenotype, including unique environment contexts, farming practices and a crop's specific growth habitat. Collectively, these contextual characteristics are said to have a character; terroir also refers to this character.

9. French Agency for ecological transition.

– the respect of animal welfare, administrative and health regulations.

These characteristics can provide a framework for reading the sustainability of the forms of urban agriculture resulting from this reconnection between the agricultural and urban worlds, and the tensions, as well as the potentialities that accompany it.

3.5. The interface between urban spaces and climate

The last of the interfaces that we are going to analyze permits us, while remaining at the scale of cities and urban spaces, to move away from them even more than through the lens of agriculture and agricultural spaces. Dealing with the interactions between urban areas and climate will thus be an opportunity to emphasize the extent to which physical phenomena interact and can be observed at different scales. It will also show how public action is organized in complex relationships with different levels of intervention, from local to global, as well as the result of pressures posed by the challenges of climate change to the dominant model of urban development in recent decades.

It is also notable that in recent years, we have seen a reversal in the way the relationship between climate and the city is conceived. Since time immemorial, knowledge on the influence of climate on urban settlements has led designers and builders to deal with it, by selecting the best sites, and adapting materials and construction processes to local climatic conditions. In recent decades, the symmetrical idea that the city influences the general climate and its own particular climate has gradually become established. However, two distinct and unequally publicized processes are at play in this effect of the city on climate: climate change and its interactions with urban systems, on the one hand, and the formation of the urban microclimate, on the other.

The climate in its multiple dimensions and scales is now recognized as a new issue – or even a motive – for land use planning. We will try to understand to what extent this recognition offers the opportunity to renew urban planning practices in the direction of improving urban sustainability and habitability.

3.5.1. *The city and climate change*

Climate change is now an undisputed fact. In order to prevent the climate system from spiraling out of control and guarantee the sustainability of the planet, the United Nations Climate Change Accord concluded in Paris in 2015 affirms the need to stabilize the increase of the global average temperature at less than 2°C – or even less than 1.5°C – by 2100, compared to pre-industrial levels[10]. The margins of maneuver to achieve this seem to be increasingly narrow. According to the scientific studies compiled by the IPCC (2018), the GHG emission trajectories allowing this objective have very strong societal implications, especially since the techniques for eliminating CO_2 from the atmosphere have not been proven on a large scale[11]. Consequently, it is urgent to massively reduce GHG emissions, which calls for profound transformations in production, consumption and lifestyles. All sectors of activity and all territories are called upon to participate in this unprecedented effort to mitigate the projected climate change. At the same time, they are or will absorb various impacts as a consequence and must now make every effort to adapt to it. In the context of the double constraint of climate risk, and faced with the rarefaction of resources, especially oil, cities, which according to UN estimates will house 68% of the world's population by 2050, are at the heart of the climate problem and its solutions.

3.5.1.1. *The contribution of cities to climate change*

Cities and urban activities (construction, thermal regulation of buildings, transport, consumption practices, etc.) are strongly involved in the dynamics of climate change, even if the precise estimation of this involvement remains difficult. In its 2011 annual report entitled "Cities and Climate Change", UN-Habitat points out the ambiguity and high imprecision that characterize the estimation of the share of anthropogenic GHG emissions attributable to cities. The production of goods and services within cities is estimated to

10. Currently, this increase is about +1°C (about +0.87°C during the decade 2006–2015 alone). Continuing at the same rate, it is expected to reach +1.5°C around 2040 and +2°C around 2060 (IPCC 2018).

11. The removal of CO_2 from the atmosphere refers to two main processes: "the enhancement of natural processes that extract carbon from the atmosphere (increased absorption by trees, soil and other 'carbon sinks') and the use of chemical processes which consist, for example, in the direct capture of CO_2 from the ambient air with a view to storing it in another medium (soil or other)" (IPCC 2018).

contribute between 40 and 60% of total anthropogenic emissions; this share rises to 60 or 70% if we take into account everything consumed by city dwellers, regardless of where it is produced (UN-Habitat 2011). This partly explains the difficulty of territorializing mitigation strategies, and above all the limits of such an exercise circumscribed to the scale of cities, the perimeter of which, as we have seen, is difficult to delimit precisely.

However, the urban targets on a global scale are well known and are mainly linked to the intense use of fossil fuels for urban transport, buildings, industrial production or waste treatment. In France, where 80% of GHG emissions are linked to fossil fuel consumption, mitigation actions at the city scale are merged with strategies to reduce this consumption. Urban energy transition strategies are therefore based, on the one hand, on improving the energy performance of urban systems (sobriety objectives) and, on the other hand, on substituting fossil fuels with renewable energy (Theys and Vidalenc 2014).

3.5.1.2. *The city: an artifact vulnerable to the effects of climate change*

In turn, cities are particularly exposed – albeit unevenly – to the effects of climate change: increases in the intensity and frequency of certain natural hazards, such as heat waves and intense precipitation; extension of urban sectors likely to be submerged in the event of storm surges due to the rise in average sea levels; modification of river regimes and flows; increases in the phenomenon of shrinkage–swelling of clays due to variations in their water content; etc. Cities are vulnerable[12] to these effects, on the one hand, due to the concentration of challenges they contend with (population, infrastructures, material goods, etc.), as well as because of the inertia of the objects and the complexity of the organizational forms which characterize them, generating interdependencies, chain reactions, domino effects, etc. Therefore, the complexity of urban systems is combined with that of the climate system to produce situations of fragility and uncertainty in which collective decision-making and actions can prove difficult. In this context, the traditional approach to risk through urban vulnerability is being challenged by the urban resilience[13] approach, which would represent a new paradigm of urban risk management in relation to climate change

12. The IPCC defines the vulnerability of a system as its propensity or susceptibility to sustain damage.

13. It is possible to define urban resilience as the capacity of an urban system to absorb a disturbance and to recover its functions after this disturbance has occurred.

(Quenault 2013). The craze for this concept, which has been perceptible in the field of urban planning since the 2000s, undoubtedly has to do with its apparent operative scope, which makes it a coveted and instrumentalized tool for local actors (Barroca et al. 2013).

In our latitudes, climate change is reflected in heat waves, or else heat waves that are predicted to become longer, more intense and more frequent. Defined by the IPCC as periods of abnormally hot weather that are difficult to tolerate, heat waves generate discomfort in both indoor and outdoor spaces of cities and pose a significant health risk to the more vulnerable populations. In this respect, the tragic heat wave of August 2003, which resulted in an estimated 15,000–20,000 excess deaths in France (70,000 on a European scale), was a wake-up call. Feedback has shown that, while individual risk factors (age, health, isolation, etc.) played a role in the observed excess mortality, the same is true for factors related to the urban environment: the hottest urban neighborhoods and within them the most poorly housed populations (poorly insulated housing, rooms directly under the roof, etc.) were the most affected (Laaidi 2012). The invitation to act on the vulnerability of cities to heat wave hazards is all the more pressing since the services of *Météo-France* ("Weather-France") announced that the summer of 2003 could correspond to a normal summer from 2070 onwards, and potentially a relatively cool summer by the end of the century. When this type of exceptional event becomes the norm, heat wave plans will no longer be sufficient; prevention must now be part of a broader approach to adapting to climate change.

Another effect of climate change is sea level rise, which threatens many cities on the most exposed coastlines and large estuaries. According to the *Institut Climate Central*[14], if the emission scenario continues business-as-usual, more than 570 coastal cities will face a sea level rise of at least 0.5 m by 2050, including major metropolises such as New York, Osaka, London, Shanghai and Jakarta. More than 800 million people in the world will be exposed to the risk of marine submersion through coastal flooding and storm surges. Actions to combat and protect urban areas by means of structures (dykes, rock filling of the coastlines, recharging sand beaches, etc.) are proving to be increasingly insufficient and temporary measures, giving way to plans for the relocation of the most sensitive urban goods and activities further inland. Jakarta, a megalopolis of 10 million inhabitants, is one of the extreme cases, due to land subsidence: the Indonesian state is preparing to

14. Climate Central Institute records data on sea level rise: Surging Seas tool.

move its capital inland to the island of Borneo, abandoning a city where about 40% of the surface area is already below sea level, and therefore at the mercy of rising waters.

In Lacanau, a seaside town in the Gironde, which emerged during the interwar period and was subsequently developed within the framework of MIACA or *Mission Interministérielle d'Aménagement de Côte Aquitaine* ("Interministerial Mission for the Development of the Aquitaine Coast", 1967–1988), the town's beach almost disappears at high tide. The coastline is receding by an average of 2 meters per year[15] and, according to experts, the retreat could reach 65 meters by 2040, and as much as 165 meters by 2100. Faced with the threat of erosion and rising waters, the municipality plans to relocate part of the waterfront inland, a neighborhood of about 1 kilometer by 250 meters. This project, which involves the demolition of 1,200 homes and 120 businesses, is estimated to cost some 300 million euros. The bill is currently out of reach for this small town of 5,000 inhabitants, not to mention the legal difficulties associated with expropriation. While waiting to be able to move the existing buildings, transitional urban planning has been called to the rescue: any new construction in the vulnerable zone must now be dismantled.

Box 3.1. *Moving the Lacanau Waterfront*

3.5.2. *Mobility and buildings: prime urban targets for mitigation*

Mitigation and adaptation are two complementary strategies for combating climate change, the former addressing the causes of the phenomenon by reducing GHG emissions, while the latter targets its consequences by attempting to limit its impacts on human communities and ecosystems. Since the rest of this chapter will focus on adaptation, the idea is to develop a few elements concerning urban mitigation strategies.

The breakdown of France's carbon footprint highlights two main emission sectors which, since the Grenelle Environment Forum, have been prime targets for urban climate policies: mobility and buildings.

The transport sector is the leading source of GHG emissions in France (30%) and increasing steadily. It is estimated that 95% of transport GHGs are due to road transport, 56% of which are attributable to private vehicles.

15. The winter storms of 2013–2014 caused significant damage to the Aquitaine coast, resulting in a severe 10-meter recession of the shoreline at Lacanau.

Outside the Paris metropolitan area[16], the modal share for the automobile is still preponderant in all urban units: 58.8% in metropolitan areas with more than 100,000 inhabitants; 66.2% in metropolitan areas with 20,000–100,000 inhabitants; and 73.2% in metropolitan areas with more than 100,000 inhabitants. Given the ecological emergency and aspirations for quality urban lifestyles, there has been a shift in urban mobility towards low-carbon modes (public transport, cycling, walking, etc.) for several years. However, health concerns related to the Covid-19 crisis are challenging these developments by reorienting travel options towards individual modes, thereby approving – alongside walking and cycling – a paradoxical "return" to the private car. Questions remain as to whether the changes underway are sustainable or transient.

The building sector is responsible for 22% of GHG emissions in France. This is due to the 3.45 billion square meters of residential and commercial buildings that are heated in winter, with some of these being air-conditioned in summer as well. Despite continuous efforts to save energy, the consumption of the sector has increased by 20% in 30 years, even if it should be noted that there is a clear progression of decarbonized energy in this evolution, at least for heating. Thanks to the constant evolution of thermal regulations, the imposition in 2006 of the DPE[17], tax incentives and direct financial aid, the energy performance of new or renovated buildings is improving significantly. However, since more than half of the primary residences date back to before the first regulation (1974), the energy renovation project remains colossal.

3.5.3. *The urban microclimate: a new challenge for urban livability*

At the local scale, each city interacts with the climate to produce a specific microclimate. This phenomenon is induced by the transformation of the geometry and nature of surfaces, which results in an alteration of energy and water exchanges between the atmosphere and the artificial urban soil. The result, at the city scale, is a modification of the wind regime, an increase in humidity and rainfall, and above all an increase in temperature as

16. The results of the SDES 2018–2019 survey are available on the webpage: https://www.statistiques.developpement-durable.gouv.fr/comment-les-francais-se-deplacent-ils-en-2019-resultats-de-lenquete-mobilite-des-personnes.

17. Diagnostic de performance énergétique: Energy Performance Diagnostic.

compared to what is observed in the surrounding rural space (Haouès-Jouve and Hidalgo 2014). This overheating effect – well known today as the urban heat island (UHI) – can reach a maximum amplitude of 2–12°C depending on the size of the area considered[18] (Oke 1982). The UHI, whose intensity is higher at night than during the day, appears as a sort of heat dome over the city, whose peak is typically at the center, corresponding to the densest and most built-up area (see Figure 3.1). It is generated by the storage of heat in materials, the reduction of evapotranspiration due to impermeable urban soils and the disappearance of vegetation, the trapping of solar radiation in the streets and a reduction of ventilation due to the buildings which form a barrier. The phenomenon of nocturnal destocking of the heat accumulated in the materials receiving solar radiation during the day leads to a rapid gradient in the nocturnal cooling of urban spaces, from the hyper-center, which is generally highly mineralized, to the zones outside the city (see Figure 3.2). In addition to these structural elements, there is the heat released by human activities, such as the heating and air conditioning of buildings, traffic or industry (anthropogenic heat), which is even more conspicuous in winter.

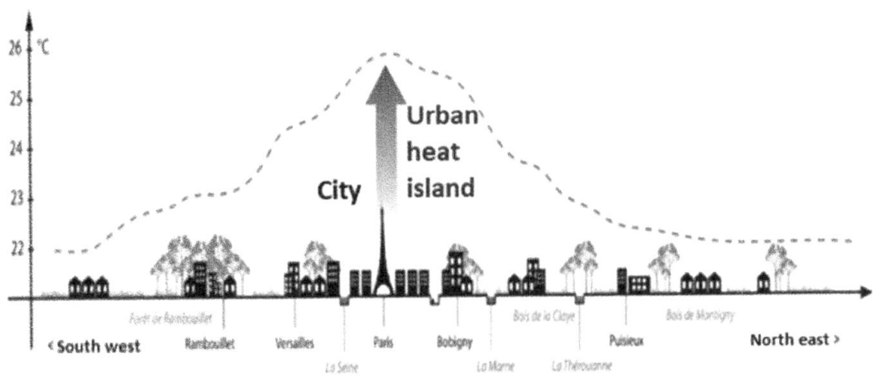

Figure 3.1. *Modeling of the urban heat island according to a south-west/north-east section of the Paris Metropolitan Area. Source: Aude Lemonsu, CNRM. For a color version of this figure, see www.iste.co.uk/adolphe/integrated.zip*

18. The UHI phenomenon tends to increase with urban development. The difference measured between the city and the surrounding countryside was estimated at +1°C in 1868 in Paris, whereas in 2003, it reached +10°C (Laaidi 2012).

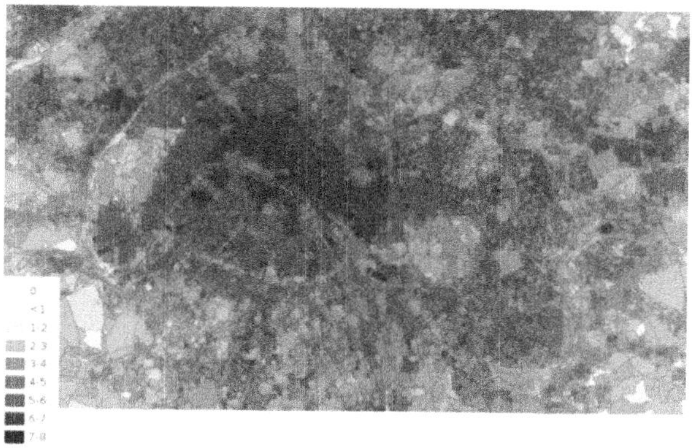

Figure 3.2. *Map of the summer night heat island (in °C) in Paris (modeled at 250 m resolution) and superimposed on building types (source: Agence d'urbanisme et d'aménagement Toulouse (AUAT) (2019)). For a color version of this figure, see www.iste.co.uk/adolphe/integrated.zip*

The UHI generates situations of discomfort or thermal stress for the inhabitants and users of urban spaces in summer. Climatic comfort or discomfort is a thermal feeling that depends on parameters specific to individuals (metabolism, activity, clothing, etc.) and the characteristics of the urban environment (air and surface temperatures, wind speed, humidity, solar radiation, etc.)[19]. During heat waves, these situations of thermal discomfort can have harmful effects on the health of the most vulnerable populations (the elderly[20], the sick, young children, etc.) and those most exposed because of their living or working conditions (poor housing, social isolation, etc.). Excessive heat limits the restorative effect of nighttime rest, promotes the formation of ozone – a toxic gas that causes irritation and respiratory difficulties – and makes it more difficult for pollutants to disperse in the air. Heat stroke, dehydration, respiratory problems, etc., are the risks most frequently associated with the peaks of excess mortality observed during heat wave episodes that are affiliated with air pollution.

The UHI is thus becoming a new factor of vulnerability for urban areas that must be controlled, especially given its effects that are only likely to

19. It is possible to assess climatic comfort using a universal indicator (UTCI: Universal Thermal Climate Index) created in 2009 by the International Society of Biometeorology.

20. The proportion of people over 75 years of age – and therefore the most vulnerable to excess heat – in the French population is expected to double by 2050 (15.6%) compared to 2003 (8.8%).

increase in the context of climate change and the intensification of heat waves (Haouès-Jouve and Hidalgo 2014). Indeed, although they stem from very distinct physical mechanisms, the formation of the urban microclimate and the dynamics of climate change can combine to increase the risks of urban overheating. The expected rise in temperatures coupled with the UHI effect is undoubtedly beneficial in winter, since it improves outdoor climatic comfort and reduces the energy demand for heating. On the other hand, it can be harmful in summer, with more or less serious consequences on the habitability of cities and the quality of urban life. In this perspective, systematic recourse to air conditioning is not a viable adaptation strategy, insofar as it deteriorates the thermal comfort of outdoor spaces and leads to an increase in energy consumption that may aggravate climate change. Furthermore, although the initial responses provided by public authorities make it possible to anticipate risk situations and manage episodic crises, particularly through heatwave plans (assistance to vulnerable persons, obligation to provide cooled rooms in hospitals and retirement homes, dissemination of behavioral advice, etc.), they are not substitute for long-term adaptation strategies.

It is becoming increasingly clear that all the actions that can be undertaken to improve the climatic comfort of indoor and outdoor spaces and thereby avoid the depreciation of the urban living environment can contribute to the mitigation of risks related to heat waves, and therefore to the climate change adaptation of the city. Several actions can be mobilized, all of which are more or less directly related to the field of urban planning and development: urban greening strategies (trees, urban and peri-urban forests, green roofs, etc.), integrating different forms of greening and encouraging the use of renewable energies. These include urban vegetation strategies (trees, urban and peri-urban forests, green roofs, etc.), including different forms of urban agriculture; the use of high-albedo materials in facades, roofs and road surfaces; the permeabilization of urban soils to facilitate the infiltration of rainwater and promote humidity; the optimization of the ventilation of urban spaces at different scales; and the protection of interior and exterior spaces from excess sunlight through better building insulation and the use of different shading techniques.

3.5.4. *Putting the climate issue on the urban policy agenda*

There are two main channels for incorporating climate issues into urban action: on the one hand, the normative channels that transmit recommendations and prescriptions vertically, from the international to the local level, via the EU and national levels; and, on the other hand, collaborative processes that disseminate and circulate expertise, know-how and good practices horizontally, within city networks, for example, or local collaborations between researchers and urban actors (Bretagne et al. 2020). For the sake of brevity, only the first channel will be discussed here.

Climate policy in France was initially structured around two founding processes: the development of the National Climate Plan – launched in 2004 and regularly updated and strengthened since – that implement the national commitments made in the framework of the Kyoto Protocol, and the "Grenelle Environment Forum" launched in 2007 to define a voluntary and consensual strategy for sustainable ecology, development and planning. Although national in scope, these processes were intended to be implemented at the local level from the outset, with an interlocking of scales and a combination of increasingly complex and extensive tools and mechanisms. The Grenelle Environment Forum and its legislative translations (the Grenelle laws of 2010) have had a significant impact in terms of integrating climate issues into urban planning and development. These laws have made it possible to modify article L101-2 of the Urban Planning Code, which now enshrines the extension of urban planning to include climate and climate change. Specific tools have also been put in place, in particular those of SCRAE[21], which are the reference documents at the regional level for policies to combat climate change, and the PCET[22], which operationalize these policies at the local level. The Grenelle forum has also led to a "greening" of general documents (SCOT[23] and PLU[24]), through which the engagement with climate change and the adaptation to its effects are becoming explicit objectives, in the same way as energy management, the effort to preserve agricultural and natural areas, or the conservation of biodiversity.

In 2015, the regulatory framework for territorial climate and energy action received a new impetus through the TECV[25] and NOTRE[26] laws.

---

21. *Schémas Régionaux Climat Air Energie*: Regional Climate Air and Energy Plans.

22. *Plans Climat Energie Territoriaux*: Territorial Climate Energy Plans.

23. *Schéma de Cohérence Territoriale*: Territorial Coherence Scheme.

24. *Plan Local d'Urbanisme*: Local Urban Plan.

25. *Transition pour la Croissance Verte*: Transition for Green Growth.

26. *Nouvelle Organisation Territoriale de la République*: New Territorial Organization of the Republic.

These laws establish the primacy of the Region/Intermunicipal combination in the definition and implementation of climate policies at the territorial level, and help to modernize the methodological arsenal at its disposal. Consequently, the SRCAEs are merging with other sectoral documents to form SRADDET[27] for new regions. The PCETs are likewise being transformed into PCAETs through the addition of air quality; but above all, they are becoming a genuinely transversal project for sustainable development, which are mandatory for any urban area of more than 20,000 inhabitants. In this renewed regulatory context, the role of cities is strengthened as the most territorialized relay for national climate policies.

However, the uneven nature of local "two-pillar" climate strategies – based on the mitigation/adaptation dualism – remains salient. Indeed, if mitigation strategies are asserted and perfected locally, it is because they manage to tie in with previous local policies and savoir-faire aimed at energy sobriety and air quality improvement (Bertrand and Richard 2015). Moreover, they are solidly framed by national standards and objectives, which are themselves variations of European Union and international commitments.

The situation is different for the "adaptation" component, which, although legally prescribed, is less understood and operationalized at the local level. Therefore, the role of local actors is often reduced – at least initially – bringing together very diverse measures and actions that are mostly "borrowed" from other sectors of urban action (the development of public spaces, risks, green spaces, tourism, etc.), in order to develop the "adaptation" component of their own climate policy (Bertrand and Richard 2015; Rocher 2013). By observing the ways in which local action on adaptation has been constructed over the past decade or so, one of the questions that arises is that of a possible "empowerment" of adaptation in relation to local climate policies that remain focused on mitigation (Rocher 2013). How should this evolution be interpreted? Should we see it as confirmation of the structural weakness of an adaptation component that is destined to remain the poor relation of local climate action, struggling to ask profound questions on the urban dynamics at work, or rather, as a positive signal for a gradual diffusion of an adaptation culture for all of urban action? For the most active cities pursuing this avenue, it would seem that the second option is emerging; the evolution outlined by Rocher concerning the

27. *Schémas Régionaux d'Aménagement, de Développement Durable et d'Egalité des Territoires*: Regional Planning, Sustainable Development and Territorial Equality Schemes.

integration of climate into Lyon's urban action seems to be confirmed: "The loss of the trace of adaptation as a structuring entry point, in terms of identified public policy, could, rather than being an abandonment, be the sign that this issue has been integrated into urban action" (Rocher 2013, author's translation).

3.5.5. *How does this translate into urban planning practices?*

Insofar as the urban microclimate in connection with climate change is becoming an issue for the preservation of well-being and health in cities, energy management, air quality and, the improvement of the urban living environment in general, it is naturally up to local authorities to take measures to integrate these issues into their actions, particularly for urban planning, development and construction. To do this, several actions can be mobilized.

Firstly, it is important to limit impermeable surfaces and vegetate urban spaces at all scales. Decades of asphalting and concreting of urban surfaces have led to the mineralization of urban soils, which reduces the cooling normally produced by the evaporation of soil moisture and the evapotranspiration of plants. It should be remembered that the limits of this type of urbanization became apparent as early as the 1970s, particularly with the worsening of flooding risks by runoff on impermeable surfaces. Since then, a doctrine has gradually developed that consists of limiting, mitigating or compensating for the impermeable urban soils so as to preserve – or recover – their ecosystem functions: habitat for biodiversity, filtration and regulation of the flow of waters that feed aquifers, elimination of contaminants, the reduction of flood and drought risks, esthetic functions, etc.[28] (European Commission 2012). In dense urban areas, soils that are kept natural or permeable[29] also help regulate the microclimate, especially when they are covered with vegetation. The latter plays a role in two ways: in

28. With the development of urban and peri-urban agriculture, these functions are supplemented by the production of foodstuffs and renewable materials such as construction timber.

29. It should be noted that awareness is slowly growing as to the importance of natural soils, the upper layer in particular, as this provides most of the ecosystem services related to soil. This awareness is based on the complexity and slowness of soil formation mechanisms (several centuries for 1 centimeter of soil). Consequently, despite the initiatives of certain cities in favor of the partial permeabilizing of sidewalks and car parks, it must be clearly understood that the mineralization of urban soils is an almost irreversible soil consumption operation.

addition to the cooling it promotes through evapotranspiration, the vegetation – when it is tall – provides shade, which promotes climatic comfort during the day. The use of vegetation on roofs and facades allows us to obtain – under certain conditions – this cooling effect in situations of scarce ground space. In all cases, the effectiveness of the cooling generated by vegetation depends on its water content, and the availability of water resources, particularly when watering may become a concern during the summer season and as such be a limiting factor.

Secondly, it is possible – at least in urban extension and renewal projects – to act on the urban form, notably the size and orientation of buildings and streets, in order to reduce the trapping of solar and infrared radiation, and facilitate natural ventilation that is conducive to heat evacuation. It is also possible to play with the thermal and radiative properties of the construction materials of buildings (facades, roofs, etc.) and the design of public spaces (pavement), in order to limit the phenomenon of heat storage due to solar radiation during the day and thus its restitution at night, in turn slowing down the cooling of the atmosphere. Materials with high albedo (light colors, low roughness, reflectivity, etc.) are particularly recommended; however, it is important to pay attention to their potential perverse effects, such as increased energy consumption for heating buildings in winter or visual discomfort generated by overly reflective surfaces (Marry 2020). It is also important to mobilize devices that increase shading and protect against excess sunlight, whether on the scale of the building (awnings, *loggias*, sunbreakers, etc.), or at the scale of public spaces. Many cities, notably in Spain and Israel, use pergolas, hanging sails and other more or less permanent shading devices to shelter streets and playgrounds from the Mediterranean sun.

Thirdly, there is growing interest in the rehabilitation of water in urban spaces, after decades of efforts to exclude it through an urban planning approach dominated by hygiene. This requires first of all the renaturation of rivers, canals, wetlands and streams, some of which have simply been channeled and buried underground. It also involves more surface management of rainwater that uses alternative techniques to the "all-pipe" policy (infiltration ditches, plot-based management, multipurpose open-air basins, etc.). On a smaller scale, there is a strong interest in devices that showcase water in public spaces, such as the water mirrors in Bordeaux, Nice and Montpellier, or the water games that are developing in most cities. At the same time, citizen movements are organizing to demand an end to the ban on bathing in public fountains that were originally designed for their esthetic appeal, or the rehabilitation and development of accessible water bodies within cities. With the increasing manifestations of climate change, urban authorities cannot remain deaf to the social demand for local cooling facilities accessible to all.

In 2017, ADEME established a three-category classification for urban cooling solutions for adapting cities to climate change:

– green and blue solutions that refer to the ecosystem services provided by nature in the city, and which can be described as nature-based solutions;

– gray solutions relating to urban infrastructures such as pavements, street furniture, materials, etc.;

– soft solutions that involve the management and uses of the city, in other words services, mobility, rhythms and lifestyles, etc.

Box 3.2. *Urban renewal and the search for co-benefits (Marry 2020)*

Within the planning exercises and urban projects that involve an integrated approach, it is interesting to combine these solutions in order to maximize the co-benefits. Indeed, at these scales, it is not so much a question of implementing specific actions for urban cooling insofar as succeeding in synergizing different actions carried out to improve well-being, the urban living environment and the environment. The "Cours Oasis" ('Schoolyard Oasis') project in Paris is a good example of this. It consists of the renovation of school and college courtyards, improving their thermal comfort by providing lighter-colored floors, more vegetation, shading devices and fountains or water game features. Conducting these projects through the prism of climate change adaptation makes it possible to converge several objectives beyond cooling alone: preservation of biodiversity; alternative management of rainwater; participation and mobilization of users; improvement of physical and mental well-being; solidarity and social ties by opening up the schoolyard to neighborhood residents during heat waves; etc.

At the scale of urban projects, it is relatively easy to imagine combining these actions to optimize climatic comfort and promote adaptation to climate change, even if these efforts often remain confined to exemplary operations, such as eco-districts, for example. At the same time, we are seeing a gradual integration of these solutions, and the climate expertise that accompanies them, into planning documents, which may constitute a way to gradual standardization. This effort is most often made within the framework of projects associating urban planning practitioners and researchers to accompany real SCOT or PLU exercises. The territorial diagnosis is the most relevant step to achieve this integration; it can contain a microclimatic diagnosis of the territory allowing for the consideration of territorialized measures for climate change adaptation. This diagnosis highlights – through the production of climatic maps of the urban environment[30] – diverse information such as the intensity and extent of nighttime UHIs, areas subject to high thermal stress during the day, as well as ventilation corridors or poorly ventilated areas. The objective is then to translate this information into the regulatory part of the document by proposing protective measures (e.g. to preserve ventilation corridors), as well as remedial or mitigating measures for the most vulnerable urban spaces (e.g. by creating cooling islands).

Although there is an international circulation of the "climate mapping" tool within a network of pioneering cities (Stuttgart, Berlin, Osaka, Hong Kong, Bilbao, Toulouse, Tunis, etc.), the methods for the rigorous treatment of climate issues in planning exercises are still fairly new and of limited application (Mhedhbi 2021). The difficulties essentially lie in understanding the phenomena that are usually dealt with at more global scales, at the local scale, which results in particular, in a lack of data and relevant analysis. It should be noted that some pioneering cities – Toulouse, Rennes, Dijon and more recently Grenoble – are accompanying their efforts to integrate climate issues into planning by setting up urban climate monitoring networks, usually in collaboration with research laboratories.

Taking into account the interactions between urban spaces and the climate is transforming the way cities are made, on the one hand, by inviting us to take a fresh look at urban materiality (site, topography, soil, atmosphere, buildings, networks, materials, etc.) and, on the other hand, by redistributing the positions of different protagonists in this process. Whether it is the result of injunctions or voluntary involvement, putting climate on the local agenda encourages cities to set new urban regulations in which the climate question, and more particularly that of adaptation, appears as an integrating issue for sectoral strategies in the areas of urban planning, energy, roads, sanitation, management of green spaces, etc.

---

30. It should be noted that the spatialization of climate data, from measurements or modeling, and their visualization in the form of climate maps, help us to facilitate adoption by urban planning practitioners.

3.6. Conclusion

The territorial approach to urban areas proposed in this chapter is obviously not exhaustive, but it does highlight some of their main characteristics, irrespective of the intrinsic diversity of urban areas around the world. The first relates to the size of the populations concerned at a time when half the world's population is urban. The second, based on the process of peripheral growth that affects most cities, refers to the areas that are now intensively or more diffusely urbanized. These two features are directly linked to a period during which mobility of all kinds has become a structuring factor, whether spatially, economically or socially. Populated, extended and connected, urban spaces are also in perpetual and sometimes accelerated evolution, some developing while others weaken.

Such developments sometimes stretch these urban features to the extreme, generating environmental and social impacts that constitute major issues to be taken into account when regulating them. The necessary regulations may seek to achieve sustainability in general, or more urgently, to improve habitability for all. However, they are no less complex to implement, given that many of the urban dynamics involved are part of an economic development model which, up until now, has not given them priority.

Beyond that, these features and developments also raise questions as to the levels of apprehension, government and governance of the urban environment, particularly in relation to such issues of habitability and sustainability. The territorial articulation corresponding to lifestyles, inscribed in economic functioning, as well as implied by the climate issue, proves to be an avenue that is just as relevant as the search for applicable perimeters for analysis and action, given that the urban has become so generalized and multifaceted.

Finally, while the reference framework of sustainable development has largely fed the observation grids of urban spaces over these last few decades, that transition is now being deployed, but is already being questioned, given that contemporary urbanization is accompanied by challenges, and even risks. Those related to climate change, in its various aspects, are widely identified, as are those related to the environmental impacts of urbanization. Those related to populations, to their inequalities in terms of housing, mobility, food, etc., always reflect this. Will urban spaces be able to prevent them? An enlightened catastrophism (Dupuy 2004) is now more necessary than ever before.

3.7. References

Agence d'urbanisme et d'aménagement Toulouse (AUAT) (2019). Urbanisme et (micro)-climat : outils et recommandations générales pour les documents de planification, issus du projet MApUCE [Online]. Available at: http://iuar-lieu-amu.fr/wp-content/uploads/2019/07/MapUCE_guide-methodologique_BD_48p.pdf [Accessed 23 March 2021].

Agier, M. (2015). *Anthropologie de la ville*. PUF, Paris.

Authier, J.-Y., Bacqué, M.-H., Guérin-Pace, F. (2007). *Le quartier. Enjeux scientifiques, actions politiques et pratiques sociales*. La Découverte, Paris.

Banos, V. and Sabatier, B. (2010). Les espaces périurbains non bâtis en France : entre publicisation "urbaine" et privatisation "rurale" ? *Articulo – Journal of Urban Research*, 3 [Online]. Available at: http://journals.openedition.org/articulo/152 [Accessed 23 March 2021].

Barles, S. (2017). Écologie territoriale et métabolisme urbain : quelques enjeux de la transition socioécologique. *Revue d'économie régionale & urbaine*, 5(5), 819–836 [Online]. Available at: https://doi.org/10.3917/reru.175.0819.

Barroca, B., Di Nardo, M., Mboumoua, I. (2013). De la vulnérabilité à la résilience : mutation ou bouleversement ? *EchoGéo*, 24 [Online]. Available at: http://journals.openedition.org/echogeo/13439 [Accessed 23 March 2021].

Beck, U. (2001). *La société du risque. Sur la voie d'une autre modernité*. Aubier, Paris.

Bertrand, F. and Richard, E. (2015). La délicate existence locale de l'adaptation aux changements climatiques : avec, sans, ou à côté de l'atténuation. *Développement durable et territoires*, 6(3) [Online]. Available at: http://developpementdurable.revues.org/11048 [Accessed 23 March 2021].

Bonin, S. (2020). Émergence en France de l'agriurbain et modèle associatif francilien : une dynamique paysagère pour les espaces périurbains ? *Territoire en mouvement, Revue de géographie et aménagement*, 44–45 [Online]. Available at: http://journals.openedition.org/tem/6186 [Accessed 19 January 2021].

Bourdin, A. (ed.) (2015). *La métropole fragile*. Éditions du Moniteur, Antony.

Bourdin, A. (2019). *Faire centre. La nouvelle problématique des centres-villes*. Édition de l'Aube, La Tour-d'Aigue.

Bretagne, G., Haouès-Jouve, S., Hidalgo, J., Touchard, O. (2020). Cartographier, modéliser. *Urbanisme*, 417, 33–34.

Burgel, G. (2003). *La ville contemporaine après 1945 – Histoire de l'Europe urbaine*, Volume 6. Le Seuil, Paris.

Castel, R. (2003). *L'insécurité sociale. Qu'est-ce qu'être protégé ?* Le Seuil et La République de la Idées, Paris.

Chabrol M., Collet A., Giroud, L.L., Rousseau M., Ter Minassian H. (2016). *Gentrifications*. Éditions Amsterdam, Paris.

Charmes, E. (2019). *La revanche des villages. Essai sur la France périurbaine*. Le Seuil, Paris.

Charmes, E. and Keil, R. (2015). The politics of post-suburban densification in Canada and France. *International Journal of Urban and Regional Research* [Online]. Available at: http://onlinelibrary.wiley.com/doi/10.1111/1468-2427.12194/full [Accessed 19 January 2021].

Charvet, J.-P. and Laureau, X. (2018). *Révolution des agricultures urbaines, des utopies aux réalités. Vers des métropoles agri-urbaines ?* Éditions France Agricole, Paris.

Commission européenne (2012). Lignes directrices concernant les meilleures pratiques pour limiter, atténuer ou compenser l'imperméabilisation des sols [Online]. Available at: https://ec.europa.eu/environment/soil/pdf/guidelines/pub/soil_fr.pdf.

Davezies, L. (2008). *La république et ses territoires : la circulation invisible des richesses*. Le Seuil/La République des Idées, Paris.

Donzelot, J. (ed.) (2012). *A quoi sert la rénovation urbaine ?* PUF, Paris.

Duchemin, E., Wegmuller, F., Legault, A. (2010). Agriculture urbaine : un outil multidimensionnel pour le développement des quartiers. *Vertigo,* 10(2) [Online]. Available at: https://id.erudit.org/iderudit/045504ar [Accessed 19 January 2021].

Dupuy, G. (2002). "Cities and automobile dependence" revisité : les contrariétés de la densité. *Revue d'économie régionale & urbaine*, 1, 141–156 [Online]. Available at: https://www.cairn.info/revue-d-economie-regionale-et-urbaine-2002-1-page-141.htm?ref=doi [Accessed 19 January 2021].

Dupuy, J.-P. (2004). *Pour un catastrophisme éclairé. Quand l'impossible est certain*. Le Seuil, Paris.

Elias, N. (2008). *La société de cours*. Champs Flammarion, Paris.

Ernwein, M. and Salomon Cavin, J. (2014). Au-delà de l'agrarisation de la ville : l'agriculture peut-elle être un outil d'aménagement urbain ? *Géocarrefour*, 89(1–2), 31–40.

Escaffre, F. (2020), Inteligencias ordinarias de lo urbano. Prácticas de los habitantes y alternativas en territorios metropolizados. Jugar, trasladarse, habitar. *BITACORA* [Online]. Available at: https://doi.org/10.15446/bitacora.v30n3.86814 [Accessed 19 January 2021].

Faburel, G. (2018). *Les métropoles barbares : démondialiser la ville, désurbaniser la terre*. Passager Clandestin, Neuvy-en-Champagne.

Ghorra-Gobin, C. (2015). *La métropolisation en question*. PUF, Paris.

GIEC (2018). Réchauffement planétaire de 1,5°C. Résumé à l'intention des décideurs, résumé technique et foire aux questions [Online]. Available at: https://www.ipcc.ch/site/assets/uploads/sites/2/2019/09/SR15_Summary_Volume_french.pdf.

Haouès-Jouve, S. and Hidalgo, J. (2014). Diminuer la vulnérabilité des villes à la hausse des températures. *Urbanisme, vulnérabilité et résilience urbaines*, 395, 48–51.

Harvey, D. (2010). *Géographie et capital. Vers un matérialisme historico-géographique*. Syllepse, Paris.

IAU (2015). Les filières courtes de proximité au sein du système alimentaire francilien [Online]. Available at: https://www.iaurif.org/nos-travaux/publications/les-filieres-courtes-de-proximite-au-sein-du-systeme-alimentaire-francilien.html [Accessed 19 January 2021].

IPCC (2018). Global Warming of 1.5°C. Special Report [Online]. Available at: https://www.ipcc.ch/sr15/ [Accessed 23 March 2021].

Kayser, B. (1997). *La renaissance rurale. Sociologie des campagnes du monde occidental*. Armand Colin, Paris.

Laaidi, K. (2012). Rôle des îlots de chaleur urbains dans la surmortalité observée pendant les vagues de chaleur. Synthèse des études sur la vague de chaleur d'août 2003. Institut de veille sanitaire, Saint-Maurice [Online]. Available at: https://www.santepubliquefrance.fr/determinants-de-sante/climat/fortes-chaleurs-canicule/documents/rapport-synthese/role-des-ilots-de-chaleur-urbains-dans-la-surmortalite-observee-pendant-les-vagues-de-chaleur.-synthese-des-etudes-realisees-par-l-institut-de-veil [Accessed 23 March 2021].

Larrère, C. (2017). *Les inégalités environnementales*. PUF/La Vie des Idées, Paris.

Le Caro, Y., Jousseaume, V., Poulot, M., Rouget, N. (2017). Agricultures et villes : des articulations renouvelées. *Annales de géographie*, 712, 553–563.

Le Gales, P. (1995). Du gouvernement des villes à la gouvernance urbaine. *Revue française de science politique*, 45e année, 1, 57–95.

Le Gales, P. (2011). *Le retour des villes européennes*. Les Presses de Sciences Po, Paris.

Lefebvre, H. (2009). *Le droit à la ville*. Economica-Anthropos, Paris.

Lorrain, D., Halpern, C., Chevauché, C. (2018). *Villes sobres. Nouveaux modèles de gestion des ressources*. Presses de Sciences Po, Paris.

Magnaghi, A. (2014). *La biorégion urbaine. Petit traité sur le territoire bien commun*. Etérotopia, Paris.

Mansfield, B. and Mendes, W. (2013). Municipal food strategies and integrated approaches to urban agriculture: Exploring three cases from the global north. *International Planning Studies*, 18(1), 37–60.

Marry, S. (eds) (2020). *Adaptation au changement climatique et projet urbain*. Parenthèse-ADEME, Marseille.

Mhedhbi, Z. (2021). Construire une expertise climatique locale pour mettre à l'agenda l'adaptation au changement climatique dans la planification et l'aménagement urbain : l'exemple du Grand Tunis. Urban Planning and Development PhD Thesis, Université Toulouse Jean Jaurès, Toulouse.

Ministere de la Transition écologique et solidaire (MTES) (2018). Biodiversité, tous vivants ! Plan biodiversité, comité interministériel biodiversité – 4 juillet 2018. Plan, MTES.

Moustier, P. and Mbaye, A. (1999). Introduction générale. In *Agriculture périurbaine en Afrique Subsaharienne*, Moustier, P., Mbaye, A., de Bon, H., Guérin, H., Pagès, J. (eds). Cirad, Montpellier.

Newman, P., Kenworthy, J., Vintila, P. (1995). Can we overcome automobile dependence? Physical planning in an age of urban cynism. *Cities*, 12(1), 53–65.

Oberti, M. and Préteceille, E. (2016). *La ségrégation urbaine*. La Découverte, Paris.

Oke, T.R. (1982). Overview of interactions between settlements and their environments. *WMO Experts Meeting on Urban and Building Climatology*, WPC-34, WMO, Geneva.

Perrin, C. (2009). Construire les campagnes méditerranéennes. Usages, aménagement et valorisations du foncier agricole périurbain en Provence et en Toscane (1950–2010). Geography Thesis, Université de Provence – Aix-Marseille I, Marseille.

Poulot, M. (2011). Des arrangements autour de l'agriculture en périurbain : du lotissement agricole au projet de territoire. Exemples franciliens. *Vertigo*, 11–12 [Online]. Available at: http://vertigo.revues.org/11188 [Accessed 15 May 2012].

Poulot, M. (2014). L'invention de l'agri-urbain en Île-de-France. Quand la ville se repense aussi autour de l'agriculture. *Géocarrefour*, 89(1,2), 11–19.

Poulot, M. (2016). Être ou ne pas être rural… Ou quand le rural se décline en ville comme à la campagne. *Pour*, 228, 69–76.

Quenault, B. (2013). Retour critique sur la mobilisation du concept de résilience en lien avec l'adaptation des systèmes urbains au changement climatique. *EchoGéo*, 24 [Online]. Available at: http://journals.openedition.org/echogeo/13403.

Rebotier, J. (2016). La fabrique des risques urbains. Agir sur les symptômes, comprendre les causes, un éclairage par le cas de Caracas (Venezuela). In *Dossier : Villes et climat : au carrefour de la recherche, de l'action et de la participation*, Haouès-Jouve, S. and Hidalgo, J. (eds). Les Cahiers des Ateliers de dialogue Recherche-Action-Expertise, Toulouse.

Rocher, L. (2013). Le climat, nouvel objet de l'action urbaine lyonnaise. In *Les territoires face aux changements climatiques. Une première génération d'initiatives locales*, Bertrand, F. and Rocher, L. (eds). Peter Lang, Brussels.

Rouillé d'Orfeuil, H., Rastoin J.-L., Chamay, A. (2015). SAT : 100 initiatives locales pour une alimentation responsable et durable. *Résolis*, March, 56.

Salomon Cavin, J. and Niwa, N. (2011). Agriculture urbaine en Suisse : au-delà des paradoxes. *Urbia : les cahiers du développement urbain durable*, 12, 3–16.

Sassen, S. (1996). *La ville globale*. Descartes et Compagnie, New York, London, Tokyo.

Sieverts, T. (2004). *Entre-ville. Une lecture de la zwischenstadt*. Éditions Parenthèses, Marseille.

Sonnino, R. (2009). Feeding the city: Towards a new research and planning agenda. *International Planning Studies*, 14(4), 425–443.

Soulard, C.-T., Margetic, C., Valette, E. (2011). Introduction : innovations et agricultures urbaines durables. *Norois*, 221, 7–10 [Online]. Available at: http://journals.openedition.org/norois/381 [Accessed 15 May 2012].

Stock, M. (2004). L'habiter comme pratique des lieux géographiques. *Espaces Temps.net* [Online]. Available at: https://www.espacestemps.net/articles/habiter-comme-pratique-des-lieux-geographiques/ [Accessed 23 March 2021].

Terrain, J.-J. (ed.) (2014). *Jardins en ville. Villes en jardin*. Éditions Parenthèse, Marseille.

Theys, J. and Vidalenc, E. (2014). Repenser les villes dans la société post-carbone. Research notebook, ADEME, Ministère de l'Écologie, du Développement durable et de l'Énergie.

Veltz, P. (2014). *Mondialisation ville et territoires. L'économie d'archipel.* PUF, Paris.

Vidal, R. and Fleury, A. (2009). La place de l'agriculture dans la métropole verte. *Projets de paysage* [Online]. Available at: http://www.projetsdepaysage.fr/fr/la_place_de_l_agriculture_dans_la_metropole_verte [Accessed 23 March 2021].

4

The City: A Space for Tensions in Modernity

Yann FERGUSON
Icam, Toulouse, France

Since 2007, more than half of the world now live in cities, as compared to only one out of ten in 1900. According to the United Nations, this ratio should be a little over 60% in 2030, that is, an urban population of 5 billion out of 8.3 billion people on Earth. This would mean there would be as many city dwellers as there were inhabitants worldwide in 1987. It seems today that "the whole world has become a city", according to Lewis Mumfort's famous formula (see *The City in History: Its Origins, Its Transformations, and Its Prospects*, 1961). In another famous article, Robert Ezra Park, one of the founders of the Chicago School to which we will return regularly, defines the city beyond the "material mechanism" and the "artificial construction". Above all, it is "something more" than an agglomeration of individuals and collective facilities, more than a constellation of institutions and administrative apparatuses: "moreover the city is a state of mind, a set of customs and traditions, attitudes and organized feelings, inherent in these customs and transmitted through traditions [...]. It is involved in the life processes of the people who compose it: it is a product of nature, and particularly, of human nature" (1990: 83, author's translation). Therefore, the "moral organization" and the "material organization" of the city interact to establish an "urban ecology", which is nothing more than the study of a living species, the human being, in its living environment.

Integrated Urban Environment Management and Resilience,
coordinated by Luc ADOLPHE. © ISTE Ltd 2022.

This living environment provides the human community with a protective framework free of natural constraints and adapted to its political, cultural and economic activities. It reveals humankind's eagerness to impose their own laws and rhythms on nature, to humanize their environment and to distance themselves from the uncertainty of the world. Le Corbusier (1887–1965), a major architect and urbanist of the 20th century, expresses this liberating and Promethean scope well in *Urbanisme* (1925): "In chaotic nature, man, for his own safety, creates an atmosphere, a zone of protection that is in agreement with what he thinks". The city is thus "the hold of the man over nature", evidence of a society that has organized itself according to the laws of reason: "Liberated, man tends towards pure geometry. It imposes that which what one calls order" (pp. 20–22, author's translation). The urbanization of the ways of life is, in this sense, a major social and anthropological fact of civilization, and as such, the modernity that derives from an exploitation of urban sedentariness.

The city is therefore both a founding and central object of sociology. Indeed, the creation of this discipline is contemporary, based on the urbanization of lifestyles and the social dynamics of modernity, of which the city is simultaneously the outcome and the vector:

– rationalization: the city illustrates the human pursuit of controlling its environment and organizing its way of life;

– individualism: the city physically brings together individuals who aspire to emancipate themselves from the collective yoke, to become the masters of their own existence;

– differentiation: the city embodies the complexification of society, which is fragmenting into self-referential subsystems.

More specifically, the city generates a so-called "urban" society that is reflected in a redefinition of the modes of sociability based on an equation that is difficult to balance: creating society, producing solidarity among emancipated, unrooted, multi-affiliated and sometimes disengaged individuals (section 4.1). This urban society is composed and decomposed according to socio-spatial settlement dynamics that reflect the stratification of society into classes guided by mechanisms of distinction. Public policies and proactive approaches aimed at building social connections are multiplying without necessarily having succeeded in reversing current trends, that is, increasing the social homogenization of urban spaces (section 4.2). The concept of the "sustainable city" is the contemporary avatar of this desire to create a society in the city, by reconciling its economic, social and environmental tensions. It nevertheless produces new forms of segregation and inequality (section 4.3).

4.1. Living together in urban societies

Due to the very principles of its material structure, the city contains unique potentialities for relationships that are built around two mechanisms: density and diversity. Density, in general, is the concentration of a large population; its housing and activities taking place within a relatively small geographical unit, allowing for the multiplication of flows and opportunities for encounters and exchanges of various kinds: goods, services, ideas, projects, etc. Diversity is the result of the initial settlement dynamics of cities, based on rural exodus and net migration, this is particularly true of American cities. Thousands of rural people or immigrants quickly become geographically concentrated, even though they have no common culture, values or history. To put this another way, cities lack the traditional ingredients of living together and must therefore mobilize other resources to generate a sense of community, essential to the realization of the common good. In the modern city, traditional community relations have been transformed to the point of becoming impersonal and anonymous. According to Emile Durkheim (1858–1917), considered the father of French sociology, industrial and urban societies are characterized by the constitution of a new regime of solidarity. In *De la division du travail social* (1893), he distinguishes two modes of social connection:

– In traditional societies, they are linked together by a "mechanical solidarity" (or solidarity based on similarity). Social cohesion is strong, but the division of labor is weak.

– In modern societies, which are based on a strong division of labor, people are linked by "organic solidarity" (or solidarity through mutual dependence: people need each other). It consists of the distribution of different roles and functions (political, economic, religious, social, etc.) among the members of society. Each one specialized in a function that makes them complementary to the others.

However, it appears that at a time when the world's population is becoming predominantly urban, urban societies are still struggling to generate solidarity in the classically agreed-upon sense: the civilized relations between people that involve a moral obligation for mutual assistance. City dwellers, largely liberated from the grip of traditions used in rural societies, become the legislators of their relational and civic commitments. If the city brings people together physically, it can at the same time distance them socially, and distance the sense for the common good. In this way, diversity, theoretically synonymous with richness, can generate social and ethnic differentiation that is more conflictual than unifying. Hence, the city is constantly reinventing itself, in order to produce an urban society that is finally united.

4.1.1. *City dwellers: liberated, but with a weakened collective conscience*

4.1.1.1. *The city: the space for a new sociability*

In *The City* (1921), Max Weber advocates that the central transformation cities underwent during the Middle Ages came about through the adoption of newly formed individual rights. Independent of the central power, cities were places for commerce, allowing their citizens to become rich, and thus able to buy their freedom from their masters. We go from a power dominated by lords and masters, to an economic power based on commercial activities. From this period was born the German proverb: "the air of the city sets you free".

In fact, in the cities, especially the larger ones, the individual is freer from constraints and the collective yoke than their counterparts in a rural village setting, where traditions and social customs determine individual destinies. The city is the place where exiled individuals live, detached from their original community, and is thus the place where those who do not belong, belong. This has repercussions on the commitment levels within relationships and in the broader community. Indeed, Georg Simmel (1958–1918) demonstrates that a specific mentality is attached to metropolitan life, making the big city the "seat of personal freedom" (Grafmeyer and Joseph 2008, p. 71, author's translation). According to him, the metropolis offers the opportunity for an immense expansion of life, through the coexistence of two worlds: a visible one and an invisible one, the latter escaping the control of others. This possibility of a secret world is linked to the individualization typical of the urban way of life: in the metropolis, the communal model is attenuated, and the possibilities and rationality of secrecy increases. Simmel contrasts the intellectualist character of the city dweller with the conservative sensibility and affectivity of the villager: "The smaller the circle that our environment forms, the more limited the borderless relations with others, and even more distressing still, is the way in which the circle watches over the individual's productions, the conduct of his life, his opinions" (Grafmeyer and Joseph 2008, p. 70, author's translation). Conversely, metropolitan urbanity produces a new, much more autonomous figure: the foreigner. The city brings together people who have been exiled from their original community. However, no one is in fact a stranger, because all are strangers: "the distance within the relationship means that near is far, but the very fact of otherness means that far is near" (Grafmeyer and Joseph 2008, pp. 53–54, author's translation). The result is an "objective", that is, the liberated city dweller who stands apart from the group because they are not rooted in the particularisms and biases of that group. This objectivity "does not indicate detachment or disinterest", but is instead derived "from the particular combination of proximity and distance, attention and indifference" (Grafmeyer and Joseph 2008, p. 55, author's translation). Simmel's words also perfectly illustrate the elective sociability that the city allows. Urban mobility frees the city dweller from spatiotemporal constraints and allows each person to nurture social ties independently of their place of residence. In other words, city dwellers are not obliged to limit their social life to their immediate vicinity; they can choose to ignore their neighbors all the while multiplying their encounters elsewhere in the city. As a "multi-belonger" who belongs to no one or any place, the city dweller remains in control of the quantity and intensity of the ties they choose to make.

In this way, Simmel distinguishes two kinds of social relations: affective relations and rational relations. The latter have taken on considerable importance in modern societies due to the multiplication of market relations. The city is a form of social organization very conducive to relations based on rationality: it is possible for individuals to live together without personally knowing one another, drowned in the anonymity of large cities, without social relations based on affectiveness holding an important place. The objectivity of rational (market) relations can be enough to ensure the integration of individuals in society, and even their personal fulfillment, since it is this type of relation that allows the individual to extract themself from the constraints that society exerts on them. However, this very freedom that allows the individual to keep their anonymity comes with a price, namely: the superficiality of human contact, the flight into extreme individualism and the withdrawal into a sociability devoid of any content. Secrecy thus generates solitude and isolation, whereas, paradoxically, the city is the space where the quantity and diversity of links are multiplied 10-fold.

The big city would thus be responsible for the "progressive dislocation of community-type social ties" (Jonas and Weidmann 2006, p. 169, author's translation). Overall, the idea is that the individual in the big city adopts a reserved mode of functioning when subjected to these ephemeral contacts. Indeed, the intensification of the "nervous life" (a Simmelian concept referring to the increasing number of possibilities for interaction in the big city), linked to the innate nature of the big city (more opportunities for encounters, etc.), can change the situation:

> If one, through the incessant contact with multitudes of people, were to answer as many internal reactions as in the small town, where one knows almost every person they have met, and where one has a positive relationship with each and every one of them, one would become completely atomized internally, and may fall into an unimaginable mental disposition [...] (Grafmeyer and Joseph 2004, pp. 67–68, author's translation).

The reserve typical of individuals within an urban environment, as Simmel calls it, is thus a regulatory mechanism "so that this pure form of interaction, which by definition has no other function than to satisfy a universal impulse to be together, at the same time imposes on each individual to stay on the sidelines of an interaction, to leave out of the exchange that which is personal" (Grafmeyer and Joseph 2004, p. 48, author's translation). Blasé-appearing individuals thus embody the modern image of the big-city dweller, tired by the multiple stimuli on offer by the metropolis, masking their feelings and not reacting to external solicitations. In the end, reserve becomes the condition for the existence of interactions, because it makes it possible to create a semblance of equality between the participants. Simmel notes that the act of speaking constitutes an end in itself, with speech serving no specific purpose other than to carry the interaction (Grafmeyer and Joseph 2008, author's translation). These exchanges in the public spaces of the city are therefore considered to be interactions of little weight, bordering on the caricature, because they are extremely codified and ritualized. The conditions for the proper functioning of these light exchanges are twofold: no one should monopolize the space, and exchanges should avoid intimate subjects: "Conversation is thus the most common and concrete manifestation of the positivity of superficial social contacts, since this superficiality is a regulator for the circulation between individuals who are supposed to be equal or who act as if they are" (Grafmeyer and Joseph 2008, p. 49, author's translation).

4.1.1.2. *The city of individuals*

In cities, especially the larger ones, the individual is thus freer from constraints and the collective yoke. These "liberated" or "emancipated" city dwellers – that is, freed from the collective yoke – claim their uniqueness. Very early on, Emile Durkheim became interested in an emerging value in the 19th century, the individual, the apparent object of a veritable cult, without this new religion succeeding to fashion institutions that would unite social actors among themselves. Individualism, an "idealized" value, appears then, to use Auguste Comte's expression, to be a "disease of the Western world". If this individualism does not succeed in maintaining a disinterested solidarity in terms of human relations, then how will people be able to live together, to associate, without conflicts tearing up the social contract?

Pierre Rosanvallon (2011) sees this image of the individual emerge during the French Revolution, when equality as a project of society is born. It is embodied in particular in a principle of equal freedom, that is, the advent of a society that is made up of independent, autonomous individuals, where the subject, responsible for themself, is able to invent their own life. This "equality-autonomy" banishes all dependencies since they imply subordination, and thus inequality in their relation with others. Contemporary capitalism, as it appeared in the 1980s, is characterized by a new mode of shaping work that values individual capacities for creation as a result of two developments. On the one hand, the very nature of production has changed: it incorporates much more scientific knowledge. On the other hand, the general conditions for the organization of production have been transformed: most of the jobs today are so specific that employees must be able to take initiatives, to respond to the unexpected. The individual is no longer identified as a component of the working class; it is now their uniqueness that has become a decisive factor of production. The result is the advent of an "individualism of uniqueness". It corresponds to the desire to reach a fully personal existence and to be the "author of one's own life".

This individuation process produces an individual whose characteristics are interlinked with the city, that which happens within the city (Jaillet 2001). For a very long time, the individual has been framed by value systems that explained or interpreted the world, whether they were religious or political ideologies, which gave meaning to their life and, in a way, assigned them one or more roles within that society. The individual was inserted in groups organized by a principle of authority or hierarchy, whether it was the family, the factory, or even the political party or union. Finally, they were established as being of a particular place: the district, the commune and the village. This produced a relatively secure individual, self-assured, with a definite place in society, certainly not always chosen and sometimes not at all, often devolved through tradition or the mere reproduction of what was. Today, we find ourselves in a completely different societal context; the systems that framed individuals are disintegrating and we find themselves more and more alone, abandoned, undecided about our identity, and even a little disenchanted. Freedom and autonomy have come with a "price to pay", and the individual is plunged into the anguish of "hyper-choice", bidden to build themself alone and to be successful. As soon as the individual is responsible for their own fate, they can no longer blame their failures on others. They are therefore condemned, in a way, to be dissatisfied, since they can never quite live up to their desires. According to Alain Ehrenberg (1998), in today's society there has been a certain "fatigue in being oneself" (author's translation). The difficulty of clinging on to something, the stipulation to learn to live with one's uncertainties, creates a strong feeling of insecurity that is characteristic of the urban condition.

4.1.1.3. *City dwellers: disengaged from the common good?*

The freedom granted by the city is accompanied with the peril of diluting any sense of belonging to a community, any allegiance to the common good and any general interest in its well-being. Indeed, sociability is intimately linked to socialization. According to Grafmeyer and Authier (2008), the concept of socialization has two different meanings: on the one hand, it refers to the learning process by which individuals internalize the values and norms of a society. On the other hand, it refers to "the various interactions that establish specific forms of relationships between individuals" (Grafmeyer and Authier 2008, p. 86, author's translation). Are urban interactions sufficient to create society? Germain and Archambault (1995) distinguish three types of sociability. The first is the so-called primary sociability, which refers to "ties involving people that define a common belonging, for example, the same family" (1995, p. 20, author's translation). Secondary sociability is distinguished by the fact that it is "much more instrumental and defined not by a community connection but by an associative connection" (p. 20, author's translation). This is the case, for example, of work relationships. Tertiary sociability, on the other hand, implies minimal ties and is defined by a "co-presence in the same place" (p. 21, author's translation). In *Déclin et survie des grandes villes américaines* (1991), Jane Jacobs notes the importance and particularity of human contacts in the street, the sum of which would be of great richness: "it constitutes at the same time a feeling of belonging to a common identity, a network of trust and mutual respect, and a possible recourse in case of personal or collective necessity" (p. 66, author's translation).

Much less convinced, the geographer Marie-Christine Jaillet (2001) questions the possibility for a society of individuals really forming a society. Indeed, the aspiration for autonomy contradicts the constraints necessary to the functioning of living together: "How can one claim autonomy – and in a certain way deny systems of interdependence – and at the same time accept the constraints that life in society implies, accepting the principle of solidarity, if there is no concrete experience of interdependence?" (p. 11, author's translation). Given that "to make society" implies the manufacture of a "collective being" integrating the "different" so as to "give itself a collective purpose and to envisage how it can exercise a solidarity between the members who compose this society" and "supposes or imposes to every individual the internalization of constraints and codes of conduct" whose acceptance by the individual will depend on a "right to solidarity" (p. 10, author's translation). Freedom in terms of relational commitments and disengagement from the common good thus gives birth to a paradox: the city is both a space where the quantity and diversity of connections are multiplied 10-fold as well as a space where people turn inward, isolate themselves and become lonely. It multiplies social links but does not generate solidarity. In 2003, for example, a heatwave caused nearly 15,000 deaths, mostly among lonely elderly people living in large cities. At the end of August 2003, the Raffarin government set up a plan to help the elderly. He proposed to devote a day of public unity towards this issue through the removal of another holiday. Whit Monday was chosen and declared a day of solidarity with the elderly in 2004. Once again, an institutional system of solidarity comes to provide for the decay of subjective solidarity.

Nonetheless, can we really deny the existence of solidarity within cities? To answer this question, Marie-Christine Jaillet (2001) distinguishes between compassionate solidarity and political solidarity, using the example of Toulouse to illustrate this. After the explosion of the AZF chemical plant on September 21, 2001, in the heat of the moment, compassionate solidarity functioned very effectively, with offers of accommodation and emergency reception, in the face of the tragedy. One month later, knowing that the neighborhoods most affected by the effects of the explosion were the working-class social-housing neighborhoods that form the approximate perimeter of the city's policy, we became aware that we would have to rehouse the populations from these neighborhoods. However, these populations are highly stigmatized and a permanent reception was not at all desired by the other districts or peri-urban cities. In other words, if compassionate solidarity generated a massive welcome for the victims, the lack of political solidarity prevented these same populations from settling permanently. For Jaillet, the weakening of political solidarity is a key issue for the future of contemporary societies. She sees a risk of disassociation when "this principle of solidarity must be applied to those of our neighbors who do not embody closeness, the similar, but are in fact the exact opposite, the image of the Other, with the greatest difference in his or her condition, culture, origin ..., with whom we must share space, cohabit" (p.12, author's translation). Indeed, today we observe behaviors of distancing from others, of refusing their proximity in housing, of refusing their presence in schools, most often against immigrants and/or the poor: "These are always described as if they embodied an irreducible difference with which there is no possible accommodation, as if they were irremediably the Other, the stranger to oneself. However, to enter in relation with somebody, it is also necessary to be persuaded that they are a little like oneself, it is necessary to recognize them as close, similar, and that requires an effort as long as there exists a social difference of condition or culture" (p. 12, author's translation). However, this tension between the diversity of the urban population and the social bond is not so very new. It is most notably founded in the very genesis of the United States of America.

4.1.2. *The diversity of the urban population: a threat to the social connection?*

4.1.2.1. *From rivalry to assimilation*

Bringing together a great diversity of traditions, cultures and nationalities, that is, inhabitants with no common normative heritage, the city often violently provokes the challenges of living together. In this regard, the United States, which has historically constructed itself and culturally accepted itself as a country of immigrants, has accumulated significant theoretical capital, the result of a century of research on the phenomena of otherness that can be associated with the migratory experience. American academics have thus formed a stream of research, urban ecology, an avenue of study that came from the Chicago School, which emphasizes the association between the urban environment and the emergence of specific forms of human character. The City of Chicago is considered as an ideal observation ground. In 1840, it had 5,000 inhabitants, in 1890, 1 million, and in 1930, 3,400,000. This meteoric growth is largely due to the interrupted arrival of migrants from many European countries. The sociologist Robert Park was particularly interested in the cycle of ethnic relations, describing the process of "disorganization-reorganization" that marks the interactions between native social groups and immigrants. Initial social disorganization refers to a reduction in the influence of social rules on individual conduct. However, according to him, this stage is only the preliminary phase in the reorganization of aptitudes and behaviors. Mutations are also initiated that allow for acculturation, a period of reorganization in the country of immigration during which an "active crossbreeding" takes place, and from which a new identity will emerge. This process follows four stages, each of which is an improvement on the previous one (de Rudder 2002; Coulon 2012):

– Rivalry (or competition) is the most basic form of interaction. This characterizes the lack of social contact between individuals, a factor that will generate conflict. The period of arrival (known territorially as the "invasion") leads to particularly intense economic rivalry for jobs and housing. He describes this phase as a "non-contact interaction" or at least as a relationship limited to the instrumental relations linked to the inevitable economic interdependence.

– Conflict is the time of the awareness of the rivalry. Rivalry is unconscious and impersonal; conflict deeply involves the individual. Individuals and groups become aware of that which opposes them is simultaneously what associates them. The conflictual stage, marked by more or less violent relations, constitutes an essential deepening of interaction, because it brings minorities into social, cultural and political life in the broadest sense, and because it forces majorities and minorities to recognize the existence of a space (in the literal as well as in the metaphorical sense) and of common stakes, and thus to think on and formalize their relations.

– Adaptation (or accommodation) is the effort that individuals and groups must make in order to adjust to social situations created by rivalry and conflict. Silent competition and antagonism are succeeded by adjustment, compromise, regulation of antagonisms through laws, adoption of new norms, ritualization, which impose themselves as a new consensus. Rivalry has not disappeared, but instead become regulated and controlled.

– Then comes assimilation, the "interpenetration and fusion" between the groups, during which differences fade and their respective values start to blend together. The latter would be an individual process, a psychological process by which individuals can emancipate themselves from their community of origin, which, during the transition period had maintained the links between the old belonging and the emerging new one. Assimilation does not imply the abandonment of all cultural differences, but their weakening and, above all, their relativization as a result of the progressive adoption of shared values and objectives.

Park's approach is therefore resolutely optimistic about the capacity of the urban environment to overcome cultural differences, to form a unifying urban order through assimilation. However, this positive process can lead to the appearance of "individual demoralization" due to difficulties in finding one's way between two systems of reference.

4.1.2.2. *Does diversity disintegrate community?*

Much more recently, another American sociologist, Robert Putnam, has revealed less optimistic results in *Bowling Alone: The Collapse and Revival of American Community* (2000). Putnam derives the title of his book from the discovery that, from 1980 to 1993, bowling club memberships declined by 40%, while the number of individual bowlers increased by 10%. He laments that Americans, while still bowling, are now bowling alone, or in pairs or threes, but no longer in teams, as they had always done. The loneliness of the bowler has become a symbol of the fragmentation of American society. Putman argues that "social capital", the networks that bind members of a society together and the norms of reciprocity and trust that flows from them, is in decline in the United States. Putman sets out to show that in the last three decades of the 20th century, three dimensions undermined the foundations of American society:

– political and civic engagement: voter turnout, political competence, political trust and political activism at the local level have declined. Americans are signing 30% fewer petitions and 40% fewer are joining consumer boycott campaigns than they were 10 or 20 years ago. The decline is also visible in non-political community life: membership and activity in all kinds of local associations, civic and religious organizations have fallen at an accelerating rate. In the mid-1970s, the average American attended several association meetings once a month; by 1998, that attendance rate had been reduced by almost 60%;

– informal social ties: in 1975, the average American had fun with friends at home 15 times a year; the equivalent figure in 1998 is half that. All leisure activities that involve doing something with someone else, playing volleyball or chamber music, are declining;

– tolerance and trust: although Americans are more tolerant of others than previous generations, they have less trust in others. Survey data provides a measure for the increase of bad faith and distrust.

This decline weakens political participation and civic engagement, and ultimately threatens the vitality of American democracy. To understand the mechanisms of this corrosion of the American social bond, Putman distinguishes between two types of social capital: "bonding capital" and "bridging capital". Bonding capital is that which is created by socialization between similar people: of the same age, race, religion, etc. He notes in particular that in the most diverse communities, trust between individuals decreases sharply. The multi-cultural society would thus lead to isolation and social anomie. To combat these effects and obtain peaceful societies in a multi-ethnic country, a second type of social capital is necessary. Bridging is what we do when we become friends with people who are not like us, such as fans of another soccer team. Putman believes that this second type of social capital is dysfunctional, based on a survey of approximately 30,000 individuals residing in 41 communities (in the urban sense of the term) spread across the United States and chosen for their heterogeneity (location, size, economic and social characteristics, etc.). Four "racial" groups were defined: Hispanics, non-Hispanic Whites, non-Hispanic Blacks and Asians. The following four findings are alarming:

– as racial diversity increases, trust between individuals decreases;

– within the most diverse communities, individuals trust their neighbors less;

– in these same communities, interracial trust is lower than elsewhere, but so is intraracial trust;

– diversity leads to anomie and social isolation.

Under further examination, it was found that racial diversity would also be responsible for lower trust in local government, lower participation in political voting, lower belief in the ability to act collectively for the common good, lower engagement in community activities, lower financial participation in charity, lower number of friends and confidants per individual, and lower measures for happiness and perceived quality of life. In fact, two phenomena increase with diversity: frequency of events and hours spent watching television. Diversity would lead to distrust, and distrust to indifference. American cities, which are among the most diverse places in the world, are thus the most affected by this disintegration of the American community. While diversity has economic benefits, according to Putman, the social cost is exorbitant. The inhabitants of the most diverse communities, he says graphically, tend to retreat into themselves, "like turtles" into their shells.

4.1.3. *Producing urban solidarity*

4.1.3.1. *Architectural and urban determinisms*

The designers of the city and its architecture are aware as to the complexity of the urban social equation, making a society from individuals. Hence, their projects generally include material determinisms intended to stimulate exchanges and social sharing. Built between 1953 and 1955 by Le Corbusier, the *Maison Radieuse* was thus thought of as a "vertical village" (Denèfle et al. 2015). The idea of a "village" refers to a very positive and somewhat mythical image of a social life that would not exist in urban areas, and even less in collective housing areas. In these common-sense representations, the village evokes a more communal life when compared to the city, where individualism and a certain indifference reign. However, the inhabitants of the *Maison Radieuse* express a certain satisfaction with the quality of the relationships they have among themselves. In fact, the individual/collective articulation was particularly taken into account by Le Corbusier to promote the development of a harmonious life. Le Corbusier seeks this balance between the individual and the collective that allows people to live close to one another, while preserving intimacy. The places of exchange – such as the hall, the park, the roof terrace, the rooms offering services or recreational areas – were designed to promote this "harmonious living". Each designed space has a corresponding function, an activity: the street for circulation, the hall as an ideal place for meetings, the roof for school and the park for relaxation. In addition to offering services and responding to needs, these spaces seem to carry a whole social life that comes on top of the simple function they were intended to fulfill. For example, the "interior streets" of the building serve about 50 dwellings. In Le Corbusier's mind, the function of this strictly pedestrian pathway was essentially circulation. However, even though these were not designed for social life, the "interior streets" allow for daily micro-exchanges that create an atmosphere that reflects the reigning spirit of the *Maison Radieuse*, because beyond the simple familiarity of faces and polite exchanges, the inhabitants recognize each other. Children's games and conversations between residents are also sheltered from the bad weather here. However, some inhabitants regret the superficiality and degradation of these connections over time. Nevertheless, even though the deterioration of the internal climate of the building is consistent with that of a broader social phenomenon, a form of resistance is at work, to the point that this slight increase in indifference towards the neighborhood is known and becomes a criterion for choosing to move in. Some people go to check on a neighbor when they have not seen them for a long time. All sorts of services are provided according to the requests of each and every person: a dog to be walked, shopping to be done in town, medicine to be picked up, a repair or DIY to be done, a tax return to be completed, and above all, the most common, the children needing to be looked after.

Other experiments do not hesitate to use deterministic urban planning and architecture, as well as precise normative devices in order to "provoke" urban solidarity. They often mobilize traditional references, fed by nostalgia for an idealized community model. This is the vision underlying the city built by the Disney company, Celebration, inaugurated in 1996 in the southwestern part of the Orlando agglomeration (Didier 1997). Celebration borrows its design from a neo-traditional model, that is, one that predates 1940, both in terms of its size (the number of inhabitants must not exceed 20,000), its architectural references (which blend styles from the past, Colonial, Victorian, etc.) and its strong sense of community. The question of recreating active participation in the community is thus at the heart of Celebration's promotion, where everything is designed to facilitate relational engagement. The houses open onto the streets thanks to oversized porches, with circulation that favors the pedestrian street, with walking elevated to the rank of the first artisan of good society between neighbors, in opposition to the automobile. A set of rules sets strict etiquette to maintain the unity of the community: obligation to mow the lawn regularly, prohibition of permanent parking in front of one's house, coloring of balconies, places to hang laundry, etc. Despite this, requests for housing have been pouring in, so much so that Disney has had to allocate them through a lottery system. However, the social and racial homogeneity of Celebration's population (94% White people, 3% Asian people, 2% Black people), mainly due to housing prices, which range from $150,000 for apartments to more than $800,000 for the largest houses, is no accident. The urban ideal pursued by Disney seems to be driven by modeling the behavior of city dwellers through strong social, economic and racial homogeneity, which is the very negation of the essence of the city: freedom and diversity. Moreover, faced with, among other problems, numerous demands from residents for a better share in decision-making, Disney handed over the management of the city to the investment fund Lexin Capital in 2004. However, the Disneyland company has continued to be involved in another urban project since 1987, the Val d'Europe, which is being developed within the framework of a public–private partnership that is unique in France, involving the State, local authorities and the Walt Disney Company. Located near Disneyland Paris, this urban and business center forms the eastern sector of the new town of Marne-la-Vallée. Celebration's urban planning is closely related to New Urbanism: the architectural styles of the buildings are neo-traditional (Haussmann, Baltard and even bear close similarity to certain London architectural styles that can also be found in western Paris). However, unlike the American city, the project does not go so far as to determine social relations. Moreover, Disney does not monopolize the governance of the project, which is the responsibility of a "decision-making triangle", with the State and an inter-municipal structure grouping together the six municipalities in the area.

More generally, producing urban sociability or "urbanity" today involves thinking about the urban composition, which must provide potential for meetings and exchanges. Urban densification (i.e. the increase in the number of inhabitants per square kilometer) reinforces the intensity of uses and tightens public spaces. The latter multiply, as do the public squares, to form centralities conducive to crossings, randomness and the mixing of heterogeneous populations. Therefore, for Grafmeyer and Authier, "public spaces represent [...] a privileged entry point for the analysis of these unstable forms of interactions that are built away from durable connections, common belonging, and shared identities" (2008, p. 93, author's translation). In the work of Isaac Joseph (1992), public space is simultaneously a space for mobility, communication and social interaction. Public space, a place of free accessibility, allows for the meeting of many individuals who share no common meaning; it is the place of civility, and allows for a meeting between all the interior and watertight worlds that make up the city. As Augustin (2010) reminds us, public space "refers to concrete sites, not merely the release or extension of the private space of the home, but evoking a set of places where co-presences between people who do not know each other can be organized" (2010, p. 79, author's translation). According to Jane Jacobs (1991), a sort of balance must be found between the privacy that individuals wish to preserve and the contacts created with the neighborhood: a distinction is necessary between the world of the city and that of private life. The majority of opportunities for contact take place in shops, and Jacobs emphasizes the crucial importance of these places in urban sociability, as the in-between places between the private and the public, where "one can linger at leisure or dash in and out" (Jacobs 1991, p. 71, author's translation).

4.1.3.2. *Re-engaging residents for the common good*

Alongside experiments drawing their references from an idealized recent past, other formulas are imagined to generate a re-engagement of inhabitants in favor of the common good. Therefore, cooperative housing makes the inhabitants actors of their housing: "The cooperative societies of inhabitants are companies with variable capital [...] which have the object to provide to their person associates the enjoyment of a housing as their main residence and to contribute to the development of their collective life [...]" (excerpts from articles L200-1 and L200-2 of the *Code la Construction et de l'Habitat*, the Construction and Housing Code). Cooperative housing responds to the housing crisis by bringing more solidarity and economic, social and environmental efficiency[1]. It allows citizens to face the housing crisis by designing together a habitat around social, ecological and economic considerations. It presents itself as an alternative for people excluded from the real estate market and offers a new type of relationship with property. It can take several forms. A project can bring together aspiring homeowners through the collective purchase of a piece of land that they will develop according to principles defined together. These criteria are often based on the search for sustainability, a unifying value that contributes, from the start, to producing a community effect. However, cooperative housing does not necessarily bring together owners in new housing. It can also bring together tenants under the aegis of a cooperative that owns the building and hands over the day-to-day management to the residents' association. This associative management implies the commitment of each person to the community, as well as numerous common spaces, a kitchen, a guest room, a workshop, a playroom or a rehearsal room, a garden or a laundry room, all of which encourage the sharing of tasks and individual responsibility.

Ownership is shared collectively through the cooperative that owns the building. The inhabitants hold shares in the cooperative. An organization in the form of a social and solidary economy offers advantages. The inhabitants take on the double quality of tenant-owner; they must therefore live on site.

---

1. See the Social and Solidarity Economy Lab, http://lelabo-ess.org/l-habitat-cooperatif-une-forme-de-logement.html [Accessed January 08, 2021].

The value of the shares they own in the cooperative is disconnected from the value of their home and it becomes impossible to speculate on its resale. A person only has one vote, regardless of the number of shares they own: each cooperator is on the same level of equality. Finally, the ownership of the real estate is collective: the cooperator does not pay more than the cost of the project. Residents' cooperatives institute a form of collective ownership. The projects are often built by the inhabitants themselves. Citizens who recognize themselves in the democratic mode of governance of cooperative housing and seek the creation of social links specific to this model. In its "Manifesto", the Utopia residents' cooperative, located in Paris, seeks:

> to value other social and human relationships on a daily basis, based on sharing and solidarity. This involves not only the collective management of common spaces, but also both spatial and social sharing. Utopia is a project that is open to its neighborhood and will offer several spaces to neighborhood actors [...]. We wish to concretely implement mutual aid and solidarity, taking into account the real disparities that exist within the group in order to lessen them through the implementation of a mode of operation that is as equitable as possible[2] (author's translation).

The cooperatives together define the characteristics of their housing (architecture, use of renewable energies, etc.) and also the spaces they wish to share. Most often, a part of the building houses a laundry room, a common guest room, a playroom for the children or a foyer open to neighborhood associations. Services can also be shared, through the implementation of childcare or car-sharing, in order to develop solidarity between neighbors and foster harmonious living. Residents' cooperatives also allow for a diffuse social mix: with residents from different social backgrounds and with disparate incomes, social mix is at the heart of the model. Cooperative housing can thus allow people with different incomes to set up a collective living project. It allows for housing maintenance by providing financially accessible housing to people from modest social backgrounds. The involvement of cooperatives, the efforts to pool together resources, share spaces, and the desire to avoid intermediaries allows rents below typical market prices. Moreover, the rent is adapted to the financial resources of each cooperator.

---

2. See the website http://coop-utop.fr/la-charte/ [Accessed January 08, 2021].

The construction of a truly participatory urban democracy is another widely debated way in which to reconnect city dwellers to community issues. For Elisa Lewis and Romain Slitine (2016), the elective representative system is outdated since it does not offer any capacity for real citizen intervention in the public debate between elections, that is, every five or six years. It is therefore profoundly out of step and totally "suboptimal" in relation to the cultural and technological possibilities of the 21st century with regard to information, mobilization and interaction. In order to imagine tools and mechanisms that would allow for the advent of a "continuous democracy", pioneers such as the *Démocratie ouverte* (Open Democracy) collective or the *Forum mondial de la démocratie* (World Democracy Forum), organized by the Council of Europe, are opening up new avenues. These democracy activists are part of the dynamics for collaborative models that make the presence of a central authority obsolete. They are multiplying *civic techs*, digital projects that deploy new tools to increase the power of citizens in political life, or to make government more transparent and collaborative. Some are working to open the doors of politics to all (parties or citizen assemblies) or to involve citizens in the legislative process, or even in the rewriting of their constitution, as is the case in Iceland. Solutions are also coming from pioneering territories where concrete utopias are being invented and where the inhabitants are being led to take their destiny in hand. Policies for "participatory democracy" have given rise to mechanisms that involve citizens in local public decisions: neighborhood councils, citizen councils, development councils or even the national commission for public debate. But here again, it is an institutional offer of participation granted from above, generating a "technocracy of participation without any real political vision or questioning of power relationships", according to the authors. Alongside these institutional arrangements, Loos-en-Gohelle (a town of 7,000 inhabitants in the Hauts-de-France region), for example, has set up a systematic consultation and co-production process. Steering committees, public meetings and questionnaires are complemented via a collaborative online citizen space, a participatory television channel (Loos TV) and a local exchange service (LES).

The social equation of urban societies is thus particularly complex: to create society, to produce solidarity between emancipated individuals, claiming their uniqueness, the possibility of secrecy, and the freedom to choose with whom and how they engage connections. For all that, they are not totally disinterested in the common good and in any collective adventure, but they no longer find themselves in the traditional forms of commitment. They illustrate "this very human characteristic of wanting to be free and committed at the same time" (Boltanski and Chiapello 1999, pp. 189–190, author's translation). Cities must also confront a social problem that is at work in urban society: the way in which the urban social fabric is organized according to social classes.

4.2. The socio-spatial dynamics of the urban population

Cities are the places where there is the greatest concentration of diverse populations, in terms of both social and cultural levels. It is the ideal melting pot for the successful production of an urban society, rich in the cohabitation of inhabitants from different origins. However, the realization of this ideal comes up against a process of socio-spatial fragmentation of cities. The most affluent inhabitants seek aggregation among their peers in order to maintain their class integrity. This results in a concentration on specific urban territories. The less well-off also group together, often wishing to do so; however, it is usually the weight of economic constraints that dictates this behavior. Zone effects are emerging, and a map of urban poverty is being constructed among neighborhoods whose inhabitants have a combination of handicaps: unemployment, lack of integration and poor housing. However, this phenomenon of precariousness also affects low-income suburban classes who, in order to live out their dream of owning a single-family home, are moving further and further away from the city center, with the further costs of daily commutes and the restoration of housing that is often of poor quality added to the debt incurred at the time of its acquisition. The city is thus unraveling, inscribing economic, social and cultural spatial differences against its iconic image as a place of all mixtures. A policy of social mixing was implemented to remake the city, that is, to recreate links between different populations. However, this strategy comes up against defensive positions on the part of certain municipalities, and its relevance is questioned, insofar as it simultaneously goes against economic, individual and societal rationalities.

4.2.1. *The fragmentation of urban society from above*

4.2.1.1. *Sociologically constrained dynamics*

This fragmentation concerns the most affluent populations. It is determined by sociological constraints, through individual or political choices. The sociologically constrained socio-spatial dynamics of the urban population concern the urban bourgeoisie, whose residential trajectories betray the search for aggregation among similar groups. Michel Pinçon and Monique Pinçon Charlot (1999 and 2007, respectively) have shed light on this process of selective self-interest by surveying the Parisian upper middle class. Indeed, the residential choices of these social classes are not determined by the price of housing. They can therefore theoretically live anywhere in the Parisian agglomeration. However, their useful residential space, the one where they statistically have a higher chance of being found, is extraordinarily narrow, which shows the weight of the sociological constraints weighing on them. They are located in a few communes in the residential west: Saint-Cloud, Boulogne, the vast, chic, unenclosed housing estates that date from the Second Empire (the new town of Le Vésinet, the parks of Maisons-Laffitte or of the château in Chatou), Versailles, Saint-Germain-en-Laye, including Neuilly, which is sometimes referred to as the 21st arrondissement of Paris. However, they are also concentrated in the western part of Paris:

– the 7th arrondissement, and in particular the Faubourg Saint-Germain, with its eponymous boulevard and square, with its three legendary cafés, *Les Deux Magots*, *Le Flore* and *Lipp*, not to mention the rue Bonaparte and La Place Saint-Sulpice;

– the 8th arrondissement with its "golden triangle" delimited by the avenues Montaigne, George-V and the Champs-Élysées;

– the 16th arrondissement with its "villages in the city", such as the villas Mulhouse, de la Réunion, Molitor or Montmorency; the Docteur-Blanche square, the Mallet-Stevens Street or the hamlet of Boulainvilliers;

– the southwest of the 17th arrondissement.

The most cherished residential wish of high society is to live among people like oneself, among one's peers, protected from intruders. The question of reproduction is essential: it is a question of avoiding misalliances, by encouraging the search for youth among themselves, in order to guarantee the greatest chances of success. As the families of nobility and the higher echelons of the bourgeoisie accumulate all forms of wealth today: "money, a lot of money, but also culture, knowledge and again relationships, an inscription in networks, all of which is condensed in the notoriety of the name". For "the dominant class has two specific forms of capital within which they themselves are transmitted and thus inherited. Patrimonial capital and worldly capital remain more or less inaccessible to those who do not benefit from them by birth" (Pinçon and Pinçon-Charlot 2007, p. 266, author's translation). The latter is a combination of different types of capital: economic, cultural, social and symbolic. The pooling of the values which each one has available to them has reinforced the accumulation of wealth: the world of the great fortunes being also that of collectors, their houses are veritable museums. By their concentration, the values of exceptional real estate reinforce each other: "one is richer among the rich", explain the two sociologists. The couple used the directories of the most selective Parisian circles to define the residential areas of high society. Members of the *Jockey Club*, the *Cercle de l'Union Interalliée*, the *Automobile-Club de France*, *Polo de Paris*, and a few other equally venerable and revered institutions are all neighbors. These clubs operate by co-option, that is, that the entry of new members is very difficult, requiring sponsorship and voting on their candidacy. The Rothschild name alone symbolizes all the other forms of wealth that are condensed in the patrimony of this well-ordered family. The arbitrary privileges linked to birth are incorporated into the ideology of blue blood, are completely inscribed in their bodies: a bourgeois, that is visible. The frantic search for the "*entre soi*" (between oneself) produces an aggregation between similar people:

> No poor so as not to hinder the landscape. However, sometimes there are a few "nouveau riche" who are still not well-developed. Never mind: the spatial segregation, which is here an aggregation of similarities, is reinforced by the creation of preserved places within these already privileged spaces. Circles are among these hyper-selective places. The seaside towns, such as Gstaad in Switzerland, are also the product of wealth combined with the awareness of belonging to an elite, concerned with managing its margins and its environment (Pinçon and Pinçon-Charlot 2007, p. 26, author's translation).

This geographic aggregation, which leads to segregation by social class or ethnic group, is true regardless of the status of the housing: property, social housing or private rental. For Éric Maurin (2011), "the concentration of wealth in only a few neighborhoods is the expression of a choice, by the most affluent and best-informed people to settle in the most stable and protected environments possible [...]. This is undoubtedly the reason why, contrary to popular belief, the poorest and most unqualified people are less concentrated in the territory than the rich" (author's translation). The neutralization of the economic constraint in residential choices thus highlights the weight of sociological constraints. To be middle-class and live in La Courneuve would be nonsense, at best a statistical aberration, and more likely a sign of decline.

4.2.1.2. *Voluntary confinement for a desired social homogeneity*

Fragmentation from above is also at work within gated communities, an urban product of American origin that consists of the fencing off of residential space in the name of a desired social homogeneity and the search for security. All over the world, residential neighborhoods are closed to unauthorized visitors. These closures take a variety of forms, ranging from armed guards posted at the entrance of an upscale residence to a crossing-type barrier restricting automobile traffic in a suburban street. The enclosed spaces are also very diverse in size and shape: from a neighborhood with several thousand inhabitants to a grouping of a few dozen people; from a group of pavilions organized around a simple private street to a complex of towers several dozen stories high with numerous facilities and amenities. However, beyond this diversity, we note everywhere a desire on the part of city dwellers to erect barriers between their residential space and the outside world (Charmes 2005).

By the early 1990s, there were 20,000 such developments in the United States, with a total population of more than eight million, initially in Arizona, California, Florida and Texas. Some 40% of new housing developments were surrounded by walls or fences. By 2001, in many US metropolises such as Houston, nearly a quarter of the homes were already located within a gated community. These residential enclaves claim to provide a living environment that endorses calm, trust, respect for rules and self-control. They are organized around collective identities that may be ethnic, cultural or even generational (such as the senior "Sun City" in Arizona). They can also rely on a religious ethos, with Megachurches, such as Megachurch South-East Christian Church, which has 22,000 members. It is possible to eat there, shop there, send your children to school, work there and go climbing without ever leaving the clearly delimited campus. Within a residential enclave, it offers city dwellers a reassuring material and emotional framework, a sociability of intimacy that contrasts sharply with the anonymous and impersonal relationships typical of urban life, far from drugs, alcohol, prostitution and violence. Through the promise of security through closure and social homogeneity, gated communities add an internal offering of services (swimming pool, golf, tennis, crèche, laundry) to conclude an "urban secession". Through this "self-closing" approach, these residents are demonstrating their distrust of politicians' ability to organize the coexistence of everyone in a communal territory, in the best interest of the general public as well as their own interests in particular. For their part, politicians have easily given in to this trend towards privatization of the city, which provides them with tax revenues and management savings (the maintenance of common facilities being provided by the community itself).

These arguments were particularly attractive to the city of Toulouse, a French city where an enclosure has reached its peak capacity. A local real estate developer made it his flagship product in the 1990s, presenting it as an innovative product for a population of young executives who had come to work in the aeronautics industry. The Toulouse model is less ambitious, but no less aimed at a relatively high-end clientele, around a name, "the closed residence", which appeals to a bourgeois imagination. The promoters proclaimed the innovative character of the product, capitalizing on the image of Toulouse, "the beloved city of new technologies", which then became the real laboratory. The target population, referred to as the "upper middle class", is young and active, highly qualified, and generally single or in couples without children. The product takes the form of residential complexes made up of one or two or even three buildings, each offering 10–20 housing units, mostly one- and two-bedroom units (40% of the housing units produced in Toulouse from 2000 onwards have one or two rooms, with 70% of tenants), often built around a swimming pool and separated from the urban roadway by wire-mesh fences (depending on the project's standing, the range of services on offer may be extended, or the landscaping may be more qualitative). The production system and the typology of the products are perfectly in line with the tax exemption procedures set up via the Méhaignerie (1986), Périssol (1996), Besson (1999) and Robien (2003) laws. These legislative measures, designed to meet the national need for housing for middle-class households, encouraged the construction of private collective residential complexes on the basis of tax-exempt investments. The closed residences thus constitute a real estate niche exploited through an offer intended for individuals seeking financial investments.

For the occupants, secure parking lots and digital entrances are presented as a response to the feeling of insecurity, a theme at the heart of the 2002 French presidential campaign: this real estate product "gives security a place like nowhere else" (according to a presentation brochure from one developer). Although these residential developments do not offer the same range of services as the American model (although a swimming pool is almost always present), and do not explicitly mobilize mechanisms of chosen social homogeneity, the security argument is central. The objective is indeed to protect oneself through technical and human systems from the "aggressions" of the city by enclosing oneself inside a dissuasive space. Today, other developers are following the Toulouse example and spreading it throughout the country. This type of urban housing thus organizes a physical closure with respect to its environment, which is materialized by barriers, automatic gates controlled remotely (from within one's home), intercoms (videophones) and entrance locks: "The objective is indeed to protect oneself through technical and human systems from the 'aggressions' of the city by enclosing oneself inside a dissuasive space" (Golovtchenko and Souchet 2005, p. 145, author's translation).

The development of entire neighborhoods based on this product has a certain number of social consequences, in particular a specialization between the center and the periphery, where the center is essentially home to single tenants or couples in collective housing without children, while the periphery is home to owners, couples with children, in single-family homes. Owners or first-time buyers occupy 65% of the principal residences in the communes of the urban area outside Toulouse and 80% within the suburban ring. On the contrary, concerns are being expressed about the capacity of neighborhoods composed entirely of closed residences to generate sociability. The term "urban secession" (Jaillet 1999) is used to describe the power relationship that closed residences seem to have with the image of the city as a public, open and supportive space. In addition, these residences, often poorly connected through impoverished public spaces, are part of a monofunctional urban fabric with no economic activity except for a few small businesses. These pitfalls are added to those more generally observed in investor products, which are stigmatized for the low quality of the buildings and their usage values, "designed by people who are not going to live there" (an elected official, quoted by Ferguson (2011), author's translation). The result is monotonous neighborhoods, strongly marked by a single architectural model. These criticisms suggest a rapid obsolescence and a medium-term desertion of private rental housing in favor of social rental housing, leading to a new zoning effect.

4.2.1.3. *Exclusive urban planning in the suburbs*

Finally, fragmentation from above can be the result of urbanization policies that betray the search to maintain a socially homogeneous settlement. Eric Charmes (2011) studies this socio-spatial dynamic in French peri-urban areas characterized by a rural landscape morphology, but whose inhabitants work in a city. This peri-urban area is a triply crumbled space, in which each rural nucleus (village, town, small city) is equipped with its own ring of suburban developments while remaining separated from the neighboring communes by an undeveloped strip of fields, meadows, parks or forests. This landscape fragmentation is coupled with a political fragmentation, linked to the maintenance of the communal level, reinforced by the laws of decentralization. Finally, the peri-urban area is socially fragmented and constitutes a socio-spatial mosaic made up of cells composed of socially homogeneous populations, sorted through the dynamics of the real estate market and residential mobility. This threefold fragmentation is the breeding ground for the transformation of municipalities into veritable "residential clubs", a phenomenon referred to by the author as "clubbization". In these "clubbed" municipalities, Eric Charmes analyzes, "the inhabitants do not behave as citizens but as consumers: they live in a municipality not as members of a political community, but as members of a club based on the shared enjoyment of a set of goods and services" (author's translation), in this case the social environment (quality of the population) and the spatial environment (maintenance of green spaces). According to this logic, the number of members of a clubbed community must be low enough and their social level high enough to avoid congestion and the destruction of environmental goods. This exclusivist tendency is particularly well explained by what the author calls the "last man in" syndrome: those who move to the peri-urban area want to be the last to arrive and oppose the construction of new housing that could destroy green spaces and the environmental quality they came for. These intentions can be seen in the urban planning regulations drawn up by the municipalities, described as "exclusive urbanism". By setting a minimum size of housing to obtain a building permit, they discriminate socially against applicants on the basis of their economic capital. By protecting their natural features, they limit construction. The preservation of landscape amenities is thus silently combined with that of the social qualities of the settlement, which essentially leads to limiting urbanization and reinforcing the social homogeneity of the settlement.

Many communes that have practiced this exclusivity are now facing its consequences, in particular the aging of their population, whose need for services increases with age (Berger et al. 2009). Their implementation in the peri-urban countryside poses a twofold problem: that of distance and therefore cost, and that of the lower density or diversity of supply. In addition, these elderly populations are generally determined to age in situ, a phenomenon known as "aging-in-place". This is explained by a strong attachment to the single-family home – linked to first-time home ownership, homeowner status and time spent in the home. There are also more emotional reasons for this choice: as a witness to the family's history and a reception area, the house opens and closes with the visits of the children and grandchildren. Aging at home, in addition to the strong emotional attachment to the home, also underlies the maintenance of independence from children, which is "culturally qualified as natural". Parents give life, they open their homes to their children, receive them, help them, and it is difficult at this stage of life to project oneself into a reverse relationship which would consist of being received by one's children.

However, peri-urban communities are most often characterized by low population density and a suburban habitat composed of single-family homes. Urban infrastructures are organized for travel by car and are not very well adapted to walking or public transport. They are particularly impractical for people with reduced mobility (insufficient lighting, almost no sidewalks, very wide streets often configured in a loop, which increases distances). With the exception of places of worship, health centers and libraries or senior citizen clubs, facilities (recreational, educational and sports) are primarily targeted at the needs of nuclear families with children or adolescents. Peri-urban areas thus constitute an objectively problematic living environment for inhabitants who have become very old and whose competence and motor skills have diminished. It would seem that, as they age and experience various life events (deaths, illness, etc.), very elderly residents of the suburbs have to re-evaluate their residential situation and their travel possibilities in light of their abilities, their material and immaterial resources and their ability to cope with the stress generated by their environment. In addition to the need to develop costly services that meet the needs of this population, "aging-in-place" presents the problem of how to renew the inhabitants of a community, other than by building new housing.

4.2.2. *The fragmentation of urban society from below*

4.2.2.1. *The imperfect aggregation of the working classes*

The socio-spatial dynamics of economically constrained urban settlement concern city dwellers whose residential trajectories are less the result of individual decisions than the impact of housing costs. It is true that the working classes may also want to live near people who have a similar way of life and preferences. However, the so-called "class communitarianism" of working-class neighborhoods appears increasingly fragile, and is not enough to appeal to the inhabitants' attachment to their neighborhood, or the city's effort to convince them of their desire to live among those like themselves. In addition to sometimes falling into the temptation of an idealized vision of the working-class world, seen as a local community with strong solidarity and "culture", it also runs the risk of being perfectly suited to the proponents of the "every man for himself" approach (Oberti 2001). The attachment of working-class households to their city is more often observed when the neighborhood has a structured history of associations and politics. The same observation can be made when we analyze requests for transfers within social housing. Requests for change by working-class people are most numerous among the most stigmatized neighborhoods.

Above all, these social categories do not have the same resources as the upper middle class, either financially or in terms of useful social knowledge and systems of relationships. So much so that, according to Pinçon and Pinçon-Charlot (1999, 2007), their aggregation is imperfect: these households cannot really choose their place of residence, which corresponds to a residence assignment through the real estate market and its prices, or due to the procedures for allocating social housing. These neighborhoods are part of the logic of the "*entre soi*" simply because there are processes of social homogenization through the sorting of populations: all those who are not accepted on the housing market at different levels end up there. In the case of the market, it is the substitute social sector, that is, the private sector housing that is in poor condition or, in the case of social housing, very distant outlying locations. The result is the appearance of veritable ghettos whose populations are victims of racial discrimination, a significant increase in unemployment and a radicalization of social inequalities. This is the case in France for sensitive urban areas, which were designated as such, in 1996, as part of the *Pacte de Relance pour la Ville* (Revival Pact for the City) and whose urban fragmentation mechanisms date back to the 1970s, the period during which the limits to large housing estates first appeared. In these areas, unemployment is at an all-time high.

In the United States, territorial separatism and social fragmentation are largely based on skin color. A Black person lives in a neighborhood populated on average by 50% Black people (even though they only represent 12.5% of the US population) and 33% White people, while a White person lives in a neighborhood populated on average by 80% White people and only 7% Black people. Segregation for Black Americans is much higher than that experienced by various ethnic groups in European cities (e.g. people from Turkey, Morocco, Pakistan, the Caribbean, etc.), while Hispanic and Asian people in the United Sates are comparable. These higher levels of Black urban segregation result from five overlapping factors (Turner and Wolman 2006). First, non-racial economic and demographic conditions explain residential patterns. Thus, for example, if a residential neighborhood differs in income levels (i.e. reflecting segregation by income) and Black people have very low incomes, then some of the racial segregation may actually reflect economic rather than racial factors. Second, the behaviors would reflect the preferences (which in a pejorative sense might be called "prejudices") of Black people and White people regarding the desired racial composition of the neighborhood in which they would like to live. Clark (1991) found that Black people generally did not want to live in neighborhoods that had more than 50% White people, while White people did not want to live in neighborhoods that had more than 20% Black people. Third, segregation is said to be the result of discriminatory (and illegal) behavior on the part of those involved in the real estate market, including members of the institutions involved in the process. Real estate agents use a variety of illegal methods to discriminate against Black housing applicants, such as the type of neighborhood or housing offered, or the information they are given about offers and the availability of financing. In the rental market, apartment owners and/or landlords may also unlawfully discriminate by not disclosing offers or, more directly, by simply choosing not to rent to Black people. Fourth, the government institutions through which the real estate allocation process operates would have the effect of encouraging or facilitating residential segregation. The institutional framework of US local governments facilitates the realization of the desire for "entre soi", more so than in European countries. Finally, public policies have a direct or indirect effect on the housing distribution process. Unlike most European countries, the United States has very little public housing (less than 2% of the US housing stock). Therefore, public housing cannot be used to combat racial segregation, especially since public housing was deliberately designed along racial lines.

A process of *gentrification* of certain Black ghettos, such as Harlem, a borough of Manhattan, is however underway (Recoquillon 2014). Long a symbol of the segregation of Black Americans and the ghettoization of their neighborhoods, Harlem has been experiencing a renaissance since the early 1980s. As early as the 1910s, Harlem became the most famous Black neighborhood and the cradle of African-American culture, especially during the period called the Harlem Renaissance (1920s), marked by a strong identity affirmation in reaction to White supremacy. In the early 1980s, a municipal commission was charged with studying the reinvestment into

Harlem, the start of its gentrification. It resulted largely in massive public investment, justified by the extent of the deterioration, but above all, to make it safer for private investment. Indeed, private investors felt that Harlem remained too risky for their investments. Federal and municipal public investments combined to also foster economic development, a central element of the "urban regeneration" rhetoric. Thus, in 1994, under the Clinton administration, Harlem was designated as an *empowerment zone* and received several hundred million dollars over 10 years to develop economic and commercial activities. However, instead of benefiting local entrepreneurs and traditional businesses, it seems that the funds mainly supported the establishment of large national chains whose competitiveness could not be matched by small stores. In an effort to protect private business interests and activities, Mayor Giuliani banned African vendors from 125th Street in 1994. These African vendors, whose booths covered the sidewalks of the main shopping street, were perceived as unfair competitors, not paying rent or taxes. A group of merchants therefore requested the intervention of the public authorities, who responded in spectacular fashion: 400 police officers in full combat gear, some on horseback, expelled the thousand vendors. However, in the end, the merchants themselves pleaded for the reintroduction of a limited number of vendors on the sidewalks of 125th Street because they are, in fact, the ones who attract customers.

4.2.2.2. *The captives of the peri-urban*

In the suburbs, behind the appealing image of social ascendancy through home ownership, significant socio-spatial inequalities are revealed. Indeed, pockets of poverty are appearing in peripheral housing estates. The poorest households are weakened by the debt associated with the purchase of a single-family home and by transportation expenses. These populations are not yet attracting public attention, which is mainly focused, in terms of territory, on emblematic areas such as so-called sensitive neighborhoods and, in terms of social issues, on the image of the homeless. It would seem that these peri-urban areas concentrate more modest households with precarious resources, but who still often receive family allowances as their only social assistance. For these "captives of a constrained location", as geographer Marie-Christine Jaillet puts it, the dream of the individual house takes the form of horizontal low-income housing, characterized by a lack of mobility, the loss of social ties and even confinement. Studying the peri-urban communes located in the "fourth ring of Toulouse", with little or no public transport and poorly equipped, Lionel Rouge (2010) shows how home ownership and settlement in these areas can be disappointing, even "painful" for working-class households, employees and even households belonging to a "small middle class" with only one salary. For these households, access to a single-family home is a sign that they belong to a social stratum, in this case the "middle" class. Their installation as homeowners is often the achievement of a lifetime of work for them.

The individuals' journeys were characterized by residential mobility with a location and type of construction explained in part by financial constraints. In addition to the debt caused by the purchase of the house, there was the need to make improvements because of the poor quality of the initial construction and the restrictions on other consumption items. The accession is thus a highly mobilizing adventure on all levels: financially, through a constrained and exaggerated recourse to credit; physically, through permanent work to improve it on a daily basis and to remedy the defects of its construction; finally, morally, through a vision proportionally positive to the negative regard towards the previous housing. However, the gap between aspirations close to those of the "middle class" and the difficulty of achieving them appears to be large, a gap which, if it does not place them in a regressive trajectory, nevertheless gives them the feeling of being blocked in their life course. Hence, Lionel Rouge speaks of "captives of the peri-urban" (author's translation). Neither the house nor the location seems to have been chosen and does not "serve" the lifestyle of the whole household. The lack of consideration weighs heavily on several elements of daily life, and in particular those of the housewife. It is instead the lifestyle of the household that finds itself in need of change to adapt to the house. This reversal is induced by a significant pressure of "preoccupations": the due date of the next repayment, the risk of unemployment, the necessity of a car trip, the tiredness, the housework to be finished and the impossibility of being able to physically detach oneself from the place of life (weekend, vacations, visits). This impossibility to realize oneself can lead to a certain frustration and bitterness, which then tends to structure the relationship with the environment, relationships and even society. There is a kind of disenchantment that tends to make daily life even more difficult than before. In these situations, the housing estate is not a symbol of social success but rather of failure. The project towards distinction fails, the means to escape are too weak and the gaze towards the other becomes unbearable. For these households, the feeling of belonging and the ability to identify are too weak. They often have the impression that they are the subject of all the changes happening, and therefore position themselves in a form of protest.

4.2.3. *The disunity of the city*

4.2.3.1. *The city is "falling apart"*

Whether sociologically (for the well-off city dwellers) or economically (for the most disadvantaged) determined, the socio-spatial dynamics of urban settlement cause the various social components to ignore each other. In France, the majority of these neighborhoods are relegated to the outskirts of the city, forming what we call the "sensitive suburbs" that are at the heart of the "urban crisis", a city that is "unraveling", as Jacques Donzelot (2006) puts it. He identifies three trends that are taking place in space and causing the city to unravel. The minorities and the poor are undergoing a process of relegation to the social housing estate, which pushes them to the outskirts of the city. The middle, small, intermediate and wealthy classes take refuge in the surrounding rural communities, which are thus urbanized and given the generic name of "peri-urban". This process of *peri-urbanization* is spreading steadily, as the poorest members of the middle classes, and then retirees, are moving further and further away, where land is cheaper and there is more peace and quiet. The city centers are affected by *gentrification*, an expression that designates the investment into the old city centers by a cultured population, anxious to have privileged access to the advantages of centrality, whether for work, pleasure or the education of their children.

These inequalities disunite urban society. The cohesion of societies depends on shared values, collective institutions and the feeling of participating in a common history and destiny. However, "the inequalities between the individuals who make up these societies can, in certain cases, erode this foundation of values and representations and, at the same time, the foundations of their cohesion", analyzes Thierry Pech (2011, author's translation). This is particularly true of societies that were built on the promise of equality and that see this horizon receding. The distance that widens between the discourses and the lived reality gradually discredits the principles and the institutions that are supposed to embody and guarantee them. The American sociologist Robert K. Merton (1938) had already considered that social disorder, *anomie* (the absence of norms), stemmed from a gap between the goals set for individuals by society and the means at the disposal of these individuals. Despite growth and the overall improvements in living conditions, inequalities of wealth have persisted and not everyone has the means to follow the consumer norms that are now prevalent in Western societies. Some people solve this contradiction through deviance.

Social cohesion is largely based on the respect of common laws that make it possible to reconcile the contradictory interests of individuals and thus limit conflicts. Therefore, it can reasonably be said that a society has a high degree of cohesion when the level of violence is low. According to this assessment criterion, the higher the degree of inequality, the lower the degree of social cohesion. In its "Global Homicide Survey" (2011), the United Nations Office on Drugs and Crime shows that "most homicides occur in countries with low levels of human development, and countries with high income inequality suffer from homicide rates that are four times higher than more equal societies". More globally, inequality is said to make societies sick, in the medical sense of the term. Richard Wilkinson and Kate Pickett (2010) highlight the link between social conditions and health: income inequality exacerbates many of a society's health and social ills (disease, drugs, delinquency, mistrust, teen pregnancy):

– the enrichment of a country is accompanied by a deterioration of its health and social situation if only the rich become richer; for the same average income, a population marked by inequality suffers significantly more than a more egalitarian population;

– inequality does not only affect the poor; the health of the better-off is also more fragile in an unequal society. In a more unequal society, there is more competition for social status and more insecurity in maintaining that status.

The authors underline the negative impact of differences in social levels within a society: if two countries have an equal average income, but one is more unequal than the other, it will have more problems in all areas: health, and also security, violence, drugs, mistrust, etc. The bottom of the social scale is certainly the most affected, but the study shows that at least 95% of a population is better off when inequalities are reduced. This would explain why, despite its status as the world's leading power, the United States has a lower life expectancy than most other developed countries (35th in the world in 2013), 710 incarcerated people out of 100,000 in the United States (compared to 123 out of 100,000 in France), more violence and more teenage births (31 births per 1,000 teenage girls, compared to 6 per 1,000 in France). Inequality appears to be the root cause of these very different evils.

4.2.3.2. *"Remaking" the city through politics: city policy in France*

The city, the solution to modernize post-war France, has in less than 50 years become the problem. The large housing estates designed in the 1950s and 1960s had a number of handicaps: economic problems (unemployment, poverty), lack of integration for migrant residents, housing shortcomings and large housing estates that never attracted city dwellers. The youth of these neighborhoods doubt that they have a future and are quick to riot, engage in illegal trafficking and depend on social assistance. The term "urban policy" appeared in the 1980s to designate a set of measures, procedures and specific practices directed towards geographic sectors defined to be of "priority", based on statistical criteria that emphasize the gap to the midpoint.

Urban policy is based on five fundamental principles. First, it affirms a desire for *solidarity* aimed at combating social and territorial differences that are deemed excessive. It then develops a *multidimensional approach* that covers housing, urban planning, education, economic development and security, but in which the housing/employment pairing is central. It is steered by a multi-partner system that systematizes the contractual framework between the State and the various communities concerned. It relies heavily on *public financing*, unlike the United Kingdom, where state investment is combined with private capital. Finally, it aims to encourage the *participation of residents* in decisions that concern them.

This city policy generates an impressive number of measures, of which we will only mention the main ones here. The prelude was the *Habitat et Vie Sociale* (Habitat and Social Life) procedure (1977), which had three objectives: to improve the comfort of tenants, to encourage the development of a harmonious social life and to foster better integration into the urban ensemble. In 1981, the *Développement Social des Quartiers* (Neighborhood Social Development) procedure, placed under the authority of the mayor, undertook to address the problems in all their educational, social, economic and preventative dimensions. At the same time, partnership initiatives were developed that took up the principle of priority geography in the action and allocation of resources: for example, this was the birth of the ZEP or *Zones d'Éducation Prioritaire* (Priority Education Zones). In 1990, for the first time, a minister in charge of the city, Michel Delebarre, was appointed, while 13 subprefects in charge of the city were also selected. A year later, a law on financial solidarity created the urban solidarity grant, paid by rich municipalities to poor municipalities in charge of a large number of low-income housing units.

In 1989, the era of contractualization began, with the introduction of city contracts. One or more local authorities together with the State agreed to jointly implement a multi-year program (initially five years, then seven years) for urban social development. From 1991 onwards, they include GPUs or *Grands Projets Urbains* (Major Urban Projects), major restructuring programs on sites in great difficulty, where the State's intervention is important. These became the *Grands Projets de Ville* (Major City Projects) in 2000. In 1996, the *Pacte de Relance* (Recovery Pact) defined a geography distinguishing between ZUS or *Zones Urbaines Sensibles* (Sensitive Urban Zones), a new label for priority neighborhoods under the city contract, ZRU or *Zones de Redynamisation Urbaine* (Urban Revitalization Zones), a subset of ZUS benefiting from tax and social security exemptions, and ZFU or *Zones Franches Urbaines* (Urban Tax-Free Zones), for neighborhoods with more than 10,000 inhabitants that are very run down. The latter benefit from additional tax and social security exemptions on the condition that one-third of their staff is employed by residents of the ZUS in the urban area.

The Solidarity and Urban Renewal Act of 2000 has left its mark on people's minds, in particular Article 55, which requires municipalities with more than 3,500 inhabitants (1,500 for the Ile-de-France region) to have a percentage of social rental housing equal to 20% of the total number of principal residences. A deduction is made from the tax resources of municipalities that do not meet the prescribed ratio. The law aims at a balanced distribution of the different types of housing, taking into account the different social categories of the inhabitants, in order to generate social diversity.

In 2002–2003, a vast program to renovate the built environment was launched, with major demolition and reconstruction operations: works to the value of 30 billion euros were carried out between 2004 and 2008, to develop public spaces, create or rehabilitate public facilities, reorganize roads, and renovate public and private housing. The State refocused procedures through the creation of the ANRU or *Agence Nationale de la Rénovation Urbaine* (National Urban Renewal Agency).

In 2008, the Fillon government announced the launch of the "Espoirs Banlieues" plan, in response to Nicolas Sarkozy's campaign promise to launch a "Marshall Plan" to "reinvent the city". Led by Fadela Amara, Secretary of State for Urban Policy (2007–2010), the plan has five pillars: employment (partnerships with business leaders for internships, etc.), education (school support, etc.), transportation (calls for projects for state aid, etc.), security (deployment of police officers in sensitive neighborhoods) and funding for local initiatives (proposed reform of state solidarity grants to local authorities).

In June 2013, François Lamy, the Minister Delegate for the City in the Ayrault government, announced a new methodology for targeting city policy credits according to the population's income. The so-called "priority neighborhoods" would be listed, that is, neighborhoods where more than half of the inhabitants live on less than 11,250 euros per year. This unique concept of "priority neighborhoods" eliminated several measures such as ZUS or ZFU. The objective was to simplify procedures and reduce the number of beneficiary neighborhoods from 2,400 to 1,300, including 10 or so "high priority" neighborhoods. In the end, 1,514 neighborhoods located in 859 municipalities benefited from the urban policy. This approach represents a turning point in relation to previous approaches, since it does not include any explicit reference to the form of the building or to difficulties relating to housing (except in certain overseas territories). It also makes it possible to concentrate resources on those who need them most. The intervention of public authorities in the 1,514 priority neighborhoods of the city policy is formalized in a partnership framework, the city contract, based on three pillars:

– social cohesion: the city contract provides for support measures to associations and social, cultural and sports facilities to promote social cohesion in the area;

– urban renewal and the living environment: the city contract funds programs for the rehabilitation or reconstruction of social housing, support for condominiums and home ownership, the construction of community facilities and the augmentation of neighborhood attractiveness;

– economic development and employment: the city contract mobilizes public employment service measures to facilitate the professional integration of neighborhood residents.

Despite these numerous measures, urban policy is not producing the expected results. The various evaluation reports all make the same observation: the Senate's report is entitled "*La politique de la ville: une efficacité entravée*" (Urban Policy: Hindered Efficiency, 2007) and the National Assembly's evaluation and control committee's report is entitled "*Quartiers défavorisés ou ghettos inavoués: la République impuissante*" (The Underprivileged Neighborhoods or Unacknowledged Ghettos: The Impotent Republic, 2010). In 2012, the *Cour des Comptes* (French National

Audit Office) established the failure of the city policy, and in particular the demolition–reconstruction operations. The report criticizes "a decade of reforms but persistent inequalities between neighborhoods", "a policy that is insufficiently managed", "an unprecedented urban renewal effort that is still poorly articulated with the social component of urban policy", "a weak mobilization of public policies in the neighborhoods" and "an unsatisfactory distribution of specific urban policy resources". More generally, urban policy is criticized for the proliferation of procedures and their entanglement, which has grown steadily over the past two decades, compiling programs, projects and agreements, all accompanied by their share of acronyms. A former senior civil servant at the Ministry of Equipment, Pierre Calame, sees its "main virtue in the creation of jobs for myriad civil servants, researchers and design offices". This lack of legibility does not make it easier to inform the citizens of the districts, whose effective participation is nevertheless desired. The second limitation stems from the difficulty of effectively implementing the desired cross-cutting approach. In other words, the various ministries contacted and the administrations concerned: National Education, Justice, Social Affairs, Interior, Equipment, Labor, etc. are used to directing the sectors they control, without much of a culture of cooperation. Contractual relations between the State and local authorities are also problematic. The latter are careful to limit any interventionism by the State. At the same time, jealous of the other's potential competencies, they cooperate poorly with each other, concerned above all to preserve the visibility of their achievements. Finally, the multiplication of intervention territories betrays the systematic problem of identifying the relevant territories for any action. President Emmanuel Macron was quick to distance himself from city policies, "this strategy is as old as I am. The first such plan was presented by Raymond Barre. It brought in some benefits, but we are at the end of what it could produce with this method". He calls for "a change of method" to get out of the "house arrest" characteristic of disadvantaged areas. Nevertheless, he has undertaken to launch various initiatives, notably on urban renewal, with a "heart of neighborhoods" operation aimed at renovating in "six months" the most severe priority neighborhoods.

4.2.3.3. *Rebuilding the city through social diversity*

Developed from the 1980s onwards, social mixing policies aim to make different social classes coexist within the same urban unit, in order to "remake" the city. The legislator thus affirms both the need to accommodate individuals in precarious or even very precarious situations in the social housing stock, and the maintenance of a diversity of profiles in order to avoid concentrations of poor people that is likely to lead to situations of residential segregation. The Solidarity and Urban Renewal Act of 2001, which requires cities of a certain size to have a 20% social housing rate, and to then increase this to 25% in municipalities with more than 3,500 inhabitants by 2025, is in accordance with this logic. Today, each urban project adheres to this objective, while taking care to avoid zoning effects, that is, segregation: social housing is dotted among the conventional housing stock in order to encourage diversity.

However, some municipalities prefer to pay the penalties, applicable to offenders, rather than apply a law that is often politically difficult to implement, particularly in well-off municipalities. Fabien Desage (2017) also shows that elected officials agree to build social housing to the extent that they can control the settlement of the population, that is, when they have the assurance of having control over the profile of future tenants. Social housing thus passes from being a constraint to a resource, insofar as it allows them to house middle-class households. Policies favoring the construction of social housing in the name of the inclusion of precarious households, as intended by the SRU law, are accompanied by a "redefinition of the target populations for who these new housing units are intended, which discreetly excludes the most undesirable populations" (Ibid, p. 24, author's translation). The most destitute are often redirected to a housing solution since, in this context of reduced supply, "the households most worthy of access must be modest enough not to be able to turn to the private sector but nevertheless able to pay rent and adopt a lifestyle that conforms to the expectations of housing actors" (Weil 2017, p. 125, author's translation).

In addition to these legal provisions, Pinçon and Pinçon-Charlot (1999, 2007) report on voluntary experiments, with a deliberate search by certain categories for a social mix presented as the best model for urban life. For example, the middle classes, in their intellectual component (teachers, technicians, social workers), cultivate a social goodwill that leads them to appreciate and seek out residential environments where it will be possible for them to rub shoulders with working-class categories. These categories cultivate "residential correctness" by looking for housing that is not too socially homogeneous and where the working classes are present in a significant way. The origins of these middle classes are often modest and a certain solidarity with these origins can thus be expressed. There is a kind of moral injunction to refuse social apartheid, to refuse anything that might resemble a kind of class racism based on political, union or association commitments. However, experience shows that spatial proximity has tended to exacerbate differences and make it difficult to live together. In Sillon de Bretagne, by placing a home for single people in difficulty amongst PLR or *Programme à Loyer Réduit* (Reduced Rent Program) housing, whose vocation was to accommodate the most disadvantaged families, and low-income housing units (HLM), a category that at the time accommodated qualified workers and employees, the construction and management organizations were facing difficulties which the entry of young households of associative activists, middle managers, teachers or doctors only aggravated further. At the time of the survey, a few years after the building was put into service, the climate was deplorable, with rumors and slander multiplying. Members of the middle classes were happy to talk about their weariness and their intention to abandon the site, the workers were dreaming of an impossible home, while the immigrant workers were the only ones to express their satisfaction at having been able to escape their previous housing conditions. There were latent conflicts over the control of the space and its sharing. This was the case for the playground, where some parents were reluctant to send their children because they considered it unsafe. This safety, which, in the discourse of some, became a leitmotif, expressing above all the daily discomfort linked to the obligatory rubbing of shoulders with certain families, all of them modest or average in condition, but with divergent dispositions and therefore expectations due to very heterogeneous histories and social trajectories. The most privileged families, when they did not leave the building, mobilized their material and social resources to move to an apartment located at the end of the longest wing, where the families belonging to the middle classes ended up regrouping. Segregation-aggregation thus found a way to reproduce itself within an explicitly anti-segregation experiment.

This principle of social diversity is therefore a challenge. According to Dominique Schnapper (2006), the equal distribution of people and their activities in space can only be a utopic vision: economic rationality and the logic of individual behavior are opposed to it. Apart from the fact that individuals have unequal economic means, which impose certain living conditions on them, how can those who are familial, socially and nationally similar not try to get closer to one another? The formal right of each person to exchange with others does not correspond to the reality of social relations. Democratic society remains heterogeneous and hierarchical; its members have different histories, capacities, social situations and solidarities. Incomes are different, as well as professions. Social relations are established within the same social circles. Depending on their social situation, individuals do not attend the same schools or the same associations, they do not practice the same sports and they do not go on vacation in the same way. The phenomena of social stratification resist spatial proximity, which is not in itself sufficient to ensure real exchanges between different social or national categories. On the contrary, it can give rise to oppositions and conflicts between socially heterogeneous individuals and groups. Proximity in space cannot be mechanically interpreted as an indicator or a factor of integration. For Donzelot (2006), the objective of social diversity is more a matter of charm than of reason. First, this is only done in one direction, that is, the reconquest of the working-class areas of the city center by the middle classes. Conversely, it is much more difficult to bring the most modest households into the beautiful neighborhoods. Second, the result is far from certain: experiments show that mixing populations does not solve the exclusion of the most disadvantaged. For Donzelot, "ghettos" do no harm anyone, as long as the residents have the means to get out of them, what he calls the "power to act", that is, the inhabitants' capacity for mobility. According to Hervé Marchal, wanting to annihilate physical borders, social separatisms or identity distinctions can be dangerous. Marchal goes so far as to quote Régis Debray and his *Eloge des frontières* (Praise of Borders): "The border is the best friend of cosmopolitanism: it recognizes that there are several worlds and that it is I who fits into them" (2010, author's translation). In other words, social mixing should not mean a renunciation of diversity, and diversity needs borders to exist. Removing territorial markers would therefore undermine the foundation of identities.

Socio-spatial dynamics make and break the urban fabric. They stir up profound social logics, which city policies struggle to reverse. For these processes, if they contradict an idealized vision of urban sociability, are paradoxically supported by certain urban aspirations in favor of the "entre soi". A new, more global approach now guides urban society, built around the principle of sustainability.

4.3. The city in search of sustainability

In 1987, the United Nations commissioned a report from Gro Harlem Brundtland, Minister of State in Norway. Entitled "Our Common Future", it is known as the Brundtland report, and popularized the notion of "sustainable development" in France. The clear definition struck peoples' minds by its universalist aspect: "Sustainable development is a development that meets the needs of the present without compromising the ability of future generations to meet their own needs" (author's translation). For the first time, an international commission stated that human activities threaten the Earth. It concludes that there is an urgency to invent a form of growth that does not penalize future generations. Chapter 8 of the report, entitled "The Urban Challenge", while mainly targeting cities in developing countries, does not fail to underline the consequences of the metropolises of industrialized countries on the environment, and, more generally, their failure to comply with the precepts of sustainable development: "These cities account for a significant share of the resources used in the world, the energy consumed and the environmental pollution produced. Many of them are able to get their resources and energy from distant lands, with disproportionate effects on the ecosystems concerned" (author's translation). Therefore, in the court of the environmental injunction, by common sense the city is accused of all evils: the inefficient use of the car, poor waste management, pollution, energy consumption, etc., and also incivility, individualism and irresponsibility. The image of the polluted city goes hand in hand with that of harmonious living. Sustainable urban development appears today as the idealized way to heal the ills of the city (Ferguson 2011). In particular, it peddles the symbolism of "the city in the country" with, in the background, the myth of the village as an idealized image of harmonious living: an urban form carrying a lost harmony between people, between people and nature. Sustainable urban development thus offers a social image to be shared, weaving together environmental considerations (the preservation of the planet), individual considerations (having a better quality of life) and collective considerations ("rebuilding" the social link).

4.3.1. *City and nature: an unlikely union*

4.3.1.1. *Between two magnets: the city in the country*

The symbol of the domestication of nature by humans, and long defined by its opposition to the countryside, the city seems to contradict the very idea of nature. In its mythology, the city imparts a high degree of domination over nature, symbolized by the *pomerium*, a sacred boundary that separates the city (*urbs*) from its surrounding territory (*ager*). When Romulus delimited the city walls of Rome by digging a furrow in the ground with a plow pulled by two oxen, he produced a rupture between the man of the cities and the man of the fields, between the creation of God (nature) and the creation of man (the circumscribed city).

The construction of this artificial world provides the human community with a protective framework free of natural constraints and adapted to its political, cultural and economic activities. The city then reveals a human capable of imposing their own laws and rhythms on nature, of humanizing their environment, and of distancing themself from the uncertainty of the world (cycles of the seasons, diseases, violence, etc.). Le Corbusier (1887–1965), a major architect and urbanist of the 20th century, expresses well in *Urbanisme* (1925), the liberating and promethean scope of the city: "In chaotic nature, man, for his own safety, creates an atmosphere, a zone of protection that is in agreement with what he thinks" (author's translation). The city is thus "the hold of man over nature"; the witness of a society which organized itself according to the laws of reason: "Free, the man tends to pure geometry. He then makes what is called order". Park in 1925 also emphasizes how much the material structure of the city weighs on representations:

> What strikes one at first sight about the city, especially the American city, is that it seems to be so little a product of the natural processes that it is difficult to see in it a living unity. The ground plan of most American cities is a checkerboard; the unit of distance is the block. This geometrical appearance suggests that the city is a purely artificial construction whose dismantling, then recomposition, can be imagined as a set of cubes (p. 86, author's translation).

In addition to this material evidence, the lifestyles of Western countries, which brought railroads to cities, have largely mobilized the private car, large contributors to cities as places where most of the air pollution and greenhouse gas emissions are concentrated. However, the designers of the city very early on sought to reconcile nature and the city.

One of the most eloquent approaches to the place of nature in the city comes from the Englishman Ebenezer Howard (1850–1928), with his "garden city", presented in *Garden Cities of To-morrow: A Peaceful Path to Real Reform* (1898). For him, it is essential to combine the advantages of city life (more active) and those of the countryside (beauty, pleasure of the fields), humans indeed attracted to being both the "city-lover" and the "country-lover": "Human society and the beauties of nature are made to be enjoyed at the same time [...]. Neither the City-Lover nor the Country-Lover fully realizes the goal of a life truly in keeping with nature. [...] It is necessary that the two magnets become one" (author's translation). His urban model is not limited to being a pleasant way of life for the rich or for the poor, but aims to achieve a synthesis of city and country.

The first major French reflections on the place of nature in the city were developed during the reign of Napoleon III. During his long exile in England, he became enthusiastic about British gardens and parks. Supported by Haussmann, he hired an engineer from the Ministry of Highways, Jean-Charles Adolphe Alphand (1817–1891). He was appointed head of the department of promenades and plantations and assumed responsibility for all the parks, promenades and squares of the Second Empire. According to the Emperor's wishes, he imposed the style of English gardens on all public green spaces: he accumulated massifs, waterfalls, rivers, grottoes and rockeries. The scope of his work made Alphand one of the greatest creators of urban green spaces in the 19th century. His creations were so enthusiastically received by his contemporaries that, unlike most of the civil servants of the Empire, he retained his position under the Third Republic, for whom he designed the gardens for the Universal Exhibition of 1889.

Although cities now house the majority of the world's population, city dwellers have not given up on the pleasures of nature. The first demand of the French where it concerns their residential environment is the presence of green spaces. Urban sprawl is the most striking form of this aspiration: preferring the individual house with a garden in a quality environment to collective housing in the heart of a dense city, city dwellers have developed huge urban areas whose ecological footprint (i.e. impact on the planet) is now being denounced. The development of peri-urban space is indeed the main feature of the development of cities. In contrast to congested and polluted city centers, this space holds the promise of greater proximity to nature, without however abandoning the pleasures of the city, which remains accessible thanks to the car. As transportation improves, the circle of the suburbs expands until it touches the old city walls. The suburbs, which had been a green area surrounding the city, now appeared as a new disease of the city. They are losing their country character and properties of isolation. They align houses that are all but identical in spaces where nature has been ransacked by the developers. It is said of these subdivisions that they often bear the names of the fauna or flora that had to be damaged in order to build them.

4.3.1.2. *The housing utopia at the heart of urban–rural contradictions*

In the post-war period, France embarked on a strategy of modernization through urban advancements. This strategy gave pride of place to theories of progressive urbanism, which praised an industrial approach to the manufacture of collective housing (see section 4.1). For the inhabitants, "it is to enter the world of hygiene, comfort, space, sun, greenery, in short, as the proponents of functionalist architecture said, it is to access paradise", says Jean-Marc Stébé (2009, author's translation). However, the realizations resulting from this architectural doctrine, in particular the large complexes, do not stand up well to the judgment of time. The addition of repetitive facades has created a monotonous urban landscape, with criticism of its gigantism, the impersonal nature of the housing, the use of low-quality materials, poor thermal and sound insulation, and social promiscuity. The "creature comforts" of this housing (central heating, bathroom, running water, indoor toilets, elevators, large bay windows, independent rooms, etc.) are not enough to save the reputation of what are commonly called "rabbit cages". Rather than progress, these dwellings have quickly become associated with unemployment, poverty and the concentration of ethnic minorities, and have become the preferred objects of urban policy. The middle classes, the primary beneficiaries of these operations, saw their residential trajectory directed by a new tropism, the individual house, a "home" that brought with it power and autonomy.

The progressive proposals were built against the suburban model. In the Charter of Athens, Le Corbusier defined the suburbs as the "domain of poor people who are tossed about by the swirls of a life without discipline" (author's translation), who on his return from the United States wrote:

> There are millions of people who want to tread the green grass of nature with their feet again; who want to see the sky, clouds and clear blue azure; who want to live with trees, those companions of the uneventful ages. Millions of them! They go, they rush, they arrive. They are now millions to see their dream murdered! Nature melts under their steps (1937, author's translation).

Born in the 19th century, the private home is considered anti-modern. Created by the bourgeoisie for the petite bourgeoisie, it is considered anti-egalitarian. More generally, the private home drew the wrath of "the France of change", since it symbolized a state of affairs that it wished to eliminate: the stability of the private house against the geographical mobility of modernity, the accumulation of the useless against the consumer society, distinction against social equality and the individualism of the pavilion against the collective personality of the greater whole.

However, the suburban home embodies the residential ideal of the middle classes. Haunted by the fear of being downgraded, the middle classes have fled social housing neighborhoods in order to better preserve a protective "entre soi" (private space) within suburban housing. The house is an ideal of life for many "middle-class" French people, and suburban housing has gradually become the residential horizon for the middle classes in western urban areas. Thanks to lower land prices than in the city center, this type of housing: the house, which evokes the farm, and the impression of returning to nature thanks to the garden, which reminds one of the fields; settling there provides a feeling of social success and constitutes an important stage in the life of the household (the purchase of the garden home often associated with the building of a family). For those who managed to escape from the working classes, it offers the unknown comfort of individual housing, a real "home", in tranquility, allowing them to "live like everyone else". Raymond et al. (2001), who investigate the "utopia of the private home", have identified several reasons for its attractiveness:

– Children: the private home offers a garden for them to expend their energy, allowing a life close to nature under easy surveillance. Indeed, the private home is an instrument of segregation against the perils of the outside world.

– Ownership: the private house gives access to the status of owner, "the culmination of a successful residential career" according to former president Nicolas Sarkozy, in 2007. The money invested in ownership is a form of asset (money invested in a property that increases in value), while the money invested in renting is synonymous with losses. This ownership creates the concept of "home", which is synonymous with autonomy and power: since it is at a distance from the neighbor, who represents a potential nuisance; thanks to self-determination of the rules, with the possibility, for example, of being able to arrange one's space freely.

– Morality: the house appears as the supreme concretization of the effort of a whole life. It must be earned, its purchase and maintenance are synonymous with hard work, far from the "ease" of renting an apartment.

– The house as a norm: the dream of access to a house reflects the normality aspiration of households. To become the owner of a house is, in a way, to make a success of one's life, especially for the middle classes of working-class origin who follow an upward trajectory thanks to their work.

The utopia of the private house has as its corollary a way of life idealized around the "village soul", which the "dense city" would prevent. The village refers to a certain way of life that favors a sense of community and civic attitudes rather than a consumerist *habitus*, which is perceived as specific to the inhabitants of large cities. The village center carries this societal ideal, this "village identity that we don't want to lose" (spoken by a mayor of the Midi-Pyrénées region, quoted by Ferguson 2014). Yves Grafmeyer speaks of a "village metaphor", to which he links four dimensions: "homogeneity of the population and lifestyles; strong identification with a small territory grouping together the essence of sociability; existence centered on the human environment of the neighborhood more than on the house; density of mutual aid networks that are also the instruments for the social control of the neighborhood" (1994, p. 80, author's translation). Charmes and Saoumi (2009) point out that

> even in the heart of large metropolises, the idea of the village is the dream. This dream has several dimensions. One of the most important has to do with sociability. If the anonymity of cities offers great freedom of action and thought, it can also be distressing. [...] The dream of the village is also the need for an anchor point in the city, the need to live in a place on a human scale, and not just a megalopolis.

This utopian housing model is gradually being taken over by public policies, which is helping to undermine the urban projects undertaken around the large housing estates. Indeed, the suburban model re-emerged as a massive phenomenon after the wave of large housing projects and priority urbanization zones (ZUP) of the 1960s (Aquachar-Charpentier 1997). Between 1970 and 1984, nearly 3 million dwellings, mostly in the form of individual housing, were produced on the outskirts of large- and medium-sized cities. The crisis context of the 1970s (economic crisis, social unrest, etc.) led the State to respond to various housing demands by initiating a policy in favor of the single-family house, known as "*aide à la pierre*" (aid for stone). These national measures were relayed by the local authorities under the effect of the 1981 decentralization law, which transferred to the mayors important competences in the field of urban planning. The latter saw in the development of peri-urban housing an effective response to the rural exodus, since this new population made it possible to revitalize trade and crafts, to prevent the closing of schools and to nourish communal finances through the housing tax. This offer meets a strong social demand from households attracted by the values of the rural environment, rejecting the progressive urbanism of the 1950s and 1970s, and seeking to create a secure heritage. This trend is reinforced by the rapid increase in the price of land.

This in turn leads to a shortage of housing in dense areas that become exacerbated in the context of strong demographic growth.

Public policies are therefore converging with the aspirations of the middle class. However, these aspirations do not imply a break with the city. The acquisition of a car is concomitant with the acquisition of a house. The households in the private house suburbs, which are often bi-motorized and described as "commuters", usually go to the city every day for work. Consequently, the urban sprawl generated by the suburban model is dependent on low gasoline prices, which was indeed the case at the time of its explosion. After the multiple crises and the emergence of an ecological conscience, this model is now highly criticized for the weight of its ecological footprint, linked to the urban sprawl it causes. The result is a growing gap between the public policies that undertake to curb the urban sprawl associated with the suburban model and the aspirations of citizens where it continues to embody the outcome of a successful residential trajectory.

4.3.1.3. *Which urban density to organize the city-nature relationship?*

Thinking about the level of density is largely influenced by the criticism on the perverse effects of peri-urban development in the form of single-family houses. Peri-urban communities fuel the dynamics of urban sprawl, described as a "global problem" by the United Nations (GEO-4 2007), and are the central issue in any strategy to reduce the ecological footprint of urban development. In France, the equivalent of a French department disappears every 10 years, causing the disappearance of agricultural land and soil sealing (Primeur 2011). From 1970 to 1984, nearly 3 million homes, mostly in the form of individual housing, were produced on the outskirts of large- and medium-sized cities, a phenomenon known as "urban exodus" (Merlin 2012). The lifestyle of these "neo-rural" people has led to individual motorization and soil mineralization, with many direct consequences: the problematic recharge of groundwater, the endangerment of natural sites and the decline of biodiversity. This model threatens nature and also weighs on the households' standard of living and public spending. For a couple, where both adults work in metropolitan France, an additional 15 kilometers away from their home can represent an additional cost of 4,000 € per year in commuting alone (Charmes and Saoumi 2009). In the United States, a study on the cost of urban sprawl estimates that the country will have to invest nearly $1,270 billion in infrastructure alone between 2000 and 2025 if urban sprawl is not kept under control and continues along its current trajectory.

The dominant idea today is therefore to remake the city on top of the city, and to make city dwellers live in compact and dense environments. This perception gained traction following the production of a curve that has made the rounds in specialized circles, the so-called "Newman and Kenworthy Curve", named after its authors, two Australian academics. They related the average density of cities to fuel consumption. We can see that two types of cities are opposed: Asian cities, which consume little energy and are very dense, and North American cities, which are very spread out and consume a lot of energy. Hence, dense urban environments, once considered as polluted, congested and dirty places, are portrayed as the model example, with dispersed housing as the threat, when they were once considered closer to nature. It is true that cities have the highest concentrations of pollution, but calculated per capita, the greenhouse gas emissions of some cities are much lower than the national average: those of New Yorkers are less than a third of the national average, and those of Barcelonans are less than half the Spanish average. Finally, there is concern about the rise in the price of land and housing, and about the possibility of housing in general for future generations. From a social point of view, the sparsely populated city would challenge the sedentary, parochial or community-based antecedents of the dense city, in which proximity was a source of connection. It is said to create an "archipelago city" in which the various blocks, often far from the main roads, tend to give rise to a kind of "urban amorphousness" (Bordreuil 2000, p. 176), that is, a disassociation between residence and traffic. This form is questioned in terms of its capacity to endow urban spaces with identity capital, since it does not contain possibilities of relational commitments for their inhabitants. It is here that an important difference sometimes emerges between elected officials, residents and professionals. Among some professionals, we note a "counter-city ideology", a long-standing condemnation that is still prevalent today, of a "spatial disorder" (Raymond et al. 2001, p. 48). Added to this is the accusation that refuses the social mixing that suburban developments, with their large plots of land, can better incorporate in a less visible manner (see section 4.2).

However, recent studies have moderated these conclusions, which are very favorable to hyperdensity. Several researchers have highlighted a phenomenon known as the "barbecue effect" (Orfeuil and Solleyret 2002). For equivalent incomes, people who live in a dense city center travel more on weekends than those who live in a suburban neighborhood. For the most affluent city dwellers, travel is sometimes by air. Density makes periods of decompression necessary, whereas the presence of a garden or green space nearby is sufficient to satisfy the need for nature on weekends. Moreover, density tends to divide public debate, stirring up a particularly stigmatizing "urban image" (Chalas 2003, p. 10): it is said to be the negation of suburban identity, everything that the inhabitants wished to escape by choosing the suburban house. It therefore seems that an intermediate density is the most appropriate: high enough to support local businesses and allow for public transportation, low enough so that city dwellers do not feel the need to go out on the weekend. A third way emerges, around the polynuclear city, in which the functions usually concentrated around the city center are dispersed among several other subcenters, forming urban nuclei or districts, linked by efficient public transport infrastructures. We thus find the Garden City of Ebenezer Howard in the background, an imagery that is very present in the new design approaches for the sustainable city.

4.3.2. *The sustainable urban project: the city as a solution*

4.3.2.1. *The city: from being a problem to being a solution*

For a long time, international institutions were reserved about urban development, but today they insist on the potentially positive character of urbanization (saving resources, possibilities of developing networks for all) as well as the need to see urban practices and policies evolve (concern for better democratic governance, combating urban sprawl, measures for promoting biodiversity in the urban context). Therefore, five years after the Brundtland Report, the Earth Summit held in Rio set up a program: "Agenda 21" (for 21st century), to move towards sustainable development. This program is based on the "Agenda 21" tool, which defines the modalities for the implementation of local initiatives for a sustainable world. The local level is considered decisive:

> Local authorities construct, operate and maintain economic, social and environmental infrastructure, oversee planning processes, establish local environmental policies and regulations, and assist in implementing national and subnational environmental policies. As the level of governance closest to the people, they play a vital role in educating, mobilizing and responding to the public to promote sustainable development

(Chapter 28, Local Government Initiatives in Support of Agenda 21)[3].

In this sense, Agenda 21 is often summarized by a formula, "*Think globally, act locally*", or by a neologism: "*glocal*".

The city thus moves from being a problem to being a solution. It is here that new behaviors and lifestyles can be invented and extended, more frugal in terms of consumption, and also a place where collective learning can take place, oriented towards the construction of a sustainable world. This is the meaning of the *Charter of European Cities and Towns Towards Sustainability*, commonly known as the *Aalborg Charter*, drafted in 1994, which begins by recognizing the ecological footprint of cities:

> We understand that our present urban lifestyle, in particular our patterns of division of labor and functions, land-use, transport, industrial production, agriculture, consumption, and leisure activities, and hence our standard of living, make us essentially responsible for many environmental problems humankind is facing[4].

The charter goes on to state that the principles of sustainable development must form the core of urban development:

> We, cities & towns, understand that the idea of sustainable development helps us to base our standard of living on the carrying capacity of nature. We seek to achieve social justice, sustainable economies, and environmental sustainability. Social justice will necessarily have to be based on economic sustainability and equity, which require environmental sustainability. Environmental sustainability means maintaining the natural capital. It demands from us that the rate at which we consume renewable material, water and energy resources does not exceed the rate at which the natural systems can replenish them, and that the rate at which we consume non-renewable resources does not exceed the rate at which sustainable renewable resources are replaced. Environmental sustainability

3. Available at: https://www.un.org/esa/dsd/agenda21/res_agenda21_28.shtml [Accessed 25th February 2022].

4. Available at: https://sustainablecities.eu/fileadmin/repository/Aalborg_Charter/Aalborg_Charter_English.pdf [Accessed 25th February 2022].

> also means that the rate of emitted pollutants does not exceed the capacity of the air, water, and soil to absorb and process them. Furthermore, environmental sustainability entails the maintenance of biodiversity; human health; as well as air, water, and soil qualities at standards sufficient to sustain human life and well-being, as well as animal and plant life, for all time.

Finally, it declares that the cities are the most strategically decisive level for initiating the necessary transformations:

> We are convinced that the city or town is both the largest unit capable of initially addressing the many urban architectural, social, economic, political, natural resource and environmental imbalances damaging our modern world and the smallest scale at which problems can be meaningfully resolved in an integrated, holistic and sustainable fashion. As each city is different, we have to find our individual ways towards sustainability.

As a result, cities are claiming a capacity for self-management and the means to implement it:

> We, cities and towns, are confident that we have the strength, the knowledge and the creative potential to develop sustainable ways of living and to design and manage our cities towards sustainability. As democratically elected representatives of our local communities we are ready to take responsibility for the task of re-organizing our cities and towns for sustainability.

Therefore, urban governance would be more inclined to develop new experiences than nation-states, which are increasingly fragmented by globalization and torn between competition and ambition. We are therefore witnessing an undeniable greening of cities, with an improvement in the quality of urban landscapes, the development of places for walking, a reflection on the permeability of soils, the greening of roofs and the return of tramways, always accompanied by a greening of the arteries they use. At the same time, experiments with sustainable neighborhoods are multiplying, under the influence of the first projects from Northern Europe in the 1990s. These "pieces of the city" are very much appreciated for their living environment: greenery, safe play areas, pedestrian and bicycle paths occupy most of the public spaces. The Hammarby Sjösdat district on the outskirts of

Stockholm and the Vauban district in the city of Freiburg illustrate the recurring principles of "eco-neighborhoods": 80% of commuting between home and work is done by bicycle or public transport, 10% of families use carpooling, water consumption is reduced by 50%, energy consumption is divided by three, and the amount of waste produced must be reduced by 20% and then by 40%. The objectives are also economic and social: the aim is to mix social categories by diversifying the housing on offer and to mix residential life with economic activity. All of this is crystallized in a structuring concept, the urban project.

4.3.2.2. *The theoretical cooling of the sustainable city*

Throughout the 20th century, the industrial production model became, for many progressive architects and urbanists, the prism through which modern societies must think of themselves, constructing their representations of the goals to be reached and the means by which to achieve them. By founding Bauhaus, Walter Gropius (1883–1969) wished to identify "the essence of a construction that is located on the common frontier of mechanics, statics, optics, acoustics, and the laws of proportion" (Gropius, quoted by Choay (1965), p.225, author's translation). Essentialism and functionality allow the emergence of the standard "unique and simplified copy of any object of use, obtained though the synthesis of the best previous forms", rid of "all the personal contribution of the designers and of all the non-essential characteristics" (Ibid., p. 226, author's translation). This standard brings to cities "that salutary homogeneity which is the true mark of a superior urban culture" (Ibid., p. 227, author's translation). It allows for the mass production of buildings, whose structure is reduced to a small number of elements. To develop the "machine city", these progressives looked to the future, cultivating an absolute belief in progress and the human rationality that must, as with industry, animate urban intentions: "We claimed in the name of the cruise liner, the airplane and the automobile, health, logic, boldness, harmony, perfection," writes Le Corbusier (1925, p. 9, author's translation). However, the achievements claiming the precepts of modern urbanism, the standardization of the built framework, the big complexes or the zoning, do not resist the judgment of the time and, with it, the faith in the techno-scientific progress that inspired them.

The society of the end of the 20th century has indeed initiated a new relationship with science, marked by a distancing of its progressive promises. The end of the growth society, the awareness of its effects on the environment and, in the background, the discrediting of scientists accused of being subservient to industrial and political interests have relegated faith in progress to the rank of an old-fashioned, naïve and even dangerous illusion. Science and technology are no longer exclusively perceived as the expression of a new mastery: "a technological innovation does not only mean a predictable technical progress, it also produces the social unknown and the unexpected environment, among others. Thus, the more modernity innovates and accentuates the movement, the more it accelerates the mutation of societies and the more it produces uncertainty" (Chalas and Soubeyran 2009, p. 11, author's translation). Knowledge remains at the center of decision-making; however, the uncertainty of society has a far more subtle link: "Science becomes increasingly necessary, but less and less sufficient for the elaboration of a socially established definition of truth" (Beck 1998, p. 344, author's translation). In other words, with societal uncertainty, the dream of clearly formulated problems and perfectly adapted and definitive solutions fades, Chalas analyzes:

> The uncertain society is not only a society lacking answers to the problems it faces. It is a society that is asking itself what are the problems that are working on it and transforming it. [...] it is not simply a society that is unable to forecast. It is a society which is reduced to discovering its harmful potentialities, especially once the beginning has taken place (2004, p. 231, author's translation).

Articulating again the thoughts of Gianni Vattimo, Chalas speaks about "weak thought" to qualify this "practical thought". They characterize periods of societal transition that "deprive thought of imperative reference points and allow it only relative orientations" (author's translation). Therefore, the State's revival policies, the model of scientific organization of work, or even doctrinal urban planning statements, are succeeded by an apprehension of problems now formulated in terms of complexity and uncertainty: without a model, without a vision of the future and without the assurance of traceable pathways, the uncertain society would make flexible systems necessary, with minimal determination.

The use of the project mode is the operational translation of the "theoretical cooling" that affects the activity of making the city. This can no longer be based on bureaucratic systems adjusted to a static and predictable vision of problems, itself associated with a repertoire of solutions, standard procedures, and the sequential and compartmentalized addition of skills: it is now a question, for each operation, of making "unidentified urban objects"

(Bourdin 2001, p. 161, author's translation). This requires cooperation between departments and the integration of new expertise in order to expand the semantic universe of that decision. While intentionality is now affected by uncertainty, the project's ambition is increasingly expressed in terms of method. It suggests a process of innovation and change, presupposing a method that recognizes "the singularity of a situation to be developed", "the exploration of opportunities in an open environment" and thus based on "the management of complexity and uncertainty" (Boutinet 1990, pp. 252–255, author's translation). In the absence of a higher reason, he refers to an empirical method in which each operation serves as an experiment to feed a broader reflection.

Without a theoretical reference point to determine choices, the reduction of uncertainty brought about by the project requires the multiplication and diversification of the actors involved in the decision-making process, as well as the integration of all the logics and interests concerned by a public decision. The challenge of this pluralist and negotiated approach is both to structure the problems in the best possible way and to find socially satisfactory answers to questions that have no undeniable technical solutions. In this perspective, the activity of making the city is increasingly developing plural, global and transversal approaches, involving all those concerned with the production, management and use of the city: politicians, administrators and technicians at central and local levels, and also economic actors and inhabitants. Municipalities are undertaking to reconfigure the systems of actors in order to reduce financial, operational and social uncertainties. First, they will largely redefine their relations with the private sector. On the one hand, they are obliged to resort to private capital – banks, real estate investors and urban service companies – to finance their operations. Financial concerns of a budgetary nature also motivate the delegation of public services to ensure a better balance of accounts. On the other hand, private operators have developed a professional offer for the development and management of public services, which leads them to position themselves as partners of project owners present at the upstream phase of defining the actions to be undertaken:

> Due to their size, their financial and human resources, and the very precise knowledge of the problems that their interventions in many sites provides, the large urban companies become co-producers of urban public action (Lorrain 1995, p. 200, author's translation).

Second, the municipalities have integrated, in a more or less intense and sincere way, the figure of the "user-actor", within the framework of participatory urban planning.

4.3.2.3. *The sustainable city: by and for its inhabitants?*

Through the lens of sustainability, the intimate relationship that inhabitants can have with the built environment and the city is perceived, particularly by urban planners and architects, as an additional vector of knowledge to be gathered upstream of the formulation of choices. Their initial involvement is also thought to promote the future appropriation of projects once they are completed. The creation of the sustainable city therefore requires a renewal of governance by developing participatory democracy (see also section 4.2). Indeed, the objectives of sustainable development and participatory democracy converge around the need to rehabilitate the commitment of city dwellers to the common good and to foster a sense of communal belonging. Participatory democracy, as a proposal for the reorganization of forms for representation and delegation, proposes new ways of thinking and acting in public action aimed at bridging the gap that has widened between the lay world (i.e. citizens) and the political and scientific worlds. Sustainable urban development is thus supposed to be achieved through "hybrid forums" (Callon et al. 2001), that is, the involvement of dialogue between citizens (bearers of vernacular knowledge), experts (bearers of techno-scientific knowledge) and elected officials (bearers of electoral legitimacy). In Vauban, an association, *le Forum Vauban*, plays multiple roles for the dispersal of information, the integration of sustainable development principles into the district, the promotion of building groups and technical support to the owners. Together, the members define the organization and the exterior layout of their future block or building. They then pass their project on to the project managers, who are appointed directly, thus saving on construction costs, as compared to traditional real estate development.

However, the participatory democracy component of the sustainable city is still far from satisfactory according to specialists. For Loïc Blondiaux, in its most common practices, it "resembles a series of communication exercises in which the real issue is more to signify the intention to involve than to actually involve" (2008, author's translation). Taoufik Saoumi also suggests that the centrality of citizens in initiating sustainable neighborhoods is a misconception. Recalling that indifference is moreover the rule in contemporary societies, he observes that in France, "sustainable neighborhoods are mainly initiated by communities and, more precisely, from a partnership formed between the technician and the politician" (2011, author's translation). Since eco-neighborhoods are very complex to implement, the integration of city dwellers throughout the process is perceived by urban leaders as an additional uncertainty rather than as a resource to reduce it.

Above all, the idea that the sustainable city is created by or for its inhabitants is fading away in the face of an empirical reality: the sustainable city primarily benefits certain inhabitants. For it is at the heart of the arguments that cities put forward in the global competition that pits them against each other, and where the major challenge is to increase their attractiveness to economic players: "For a company that is looking to set up in Europe, the choice is made not so much at the level of States as at the level of cities and territories. You generally have three or four major European cities competing with each other" (author's translation), explained Gérard Collomb, the former mayor of Lyon. Like Silicon Valley, territories are tending to become "competitiveness clusters" within the globalized world. At the same time, cities must increase their residential attractiveness to attract the "creative class", an elite that is attentive to the quality of its living environment, the brand image of its environment or the reputation of its city, and the personal or professional opportunities that the big city offers. In this sense, the sustainable city is undeniably part of a global urban marketing strategy targeting firms and their executives. This creates close links between sustainability and gentrification (Lees 2003). By participating in the production of carefully differentiated social spaces capable of adapting to the increased diversity of urban lifestyles, and by contributing to the affirmation of a socially selective urbanity, the sustainable city constitutes a new stage in the gentrification of certain neighborhoods. The environment, long excluded from the speculative processes that fuel the production of urban spaces, now seems to be becoming one of their preferred resources, with the affirmation of eco-gentrification dynamics (Béal et al. 2011).

Finally, the sustainable city is socially discriminatory and also generates gender inequalities. Yves Raibaud (2015) shows that consensual solutions (development of two-wheelers, walking, public transport, carpooling, etc.) are also those that increase inequalities between men and women. Women, of all ages, would be disadvantaged by the "good practices" of mobility in the sustainable city, and in particular the abandonment of the car. The reasons for this are both the nature of the tasks that are still mainly assigned to them (accompanying children, the elderly, shopping) and their feeling of insecurity in public space (fear of aggression in certain districts or at night). Women are much less likely to cycle (especially at night or when it is raining) and they all but abandon it if they have more than one child. Harassment in the street and on public transport appears, although there is very little anecdotal evidence, to be so systematic (100% of women have been victims, according to the 2015 report of the High Council for Equality between Women and Men) that one is surprised by the taboo surrounding this subject, which is nevertheless central to the implementation of alternative mobilities. Under these conditions, the car, more than a mobility tool, represents a means of protection to face the night. In his work on the Bordeaux metropolis through a consultation on new urban mobility, Yves Raibaud (2014) notes how the concerns expressed by women (particularly those concerning children, the elderly or the disabled and safety) are ignored or judged as "special cases", and thus left out of the conclusions and summaries of the sessions in favor of subjects that seem more important in the eyes of men: the creative, intelligent, post-carbon, hyperconnected city: "Many aspects of women's daily lives are therefore downplayed, returned to private life: how dare we say that we need the car to accompany the children or that we are afraid to walk in the city at night when it is a question of the future of the planet and the greater good?" (author's translation). The promise of a sustainable city that is quiet, healthier, recreational and conducive to living together requires that everyone make an effort to adapt to it, but in reality, "the new practices that result from it resemble as closely as possible the practices of young men, free of family obligations and in good health" (author's translation). In the sustainable city, the reproduction of gender inequalities is therefore carried out under a democratic guise. Meanwhile, the daily management of energy savings, waste, food and health continues to fall mainly to women: "This legitimizes the expression of a critical ecofeminism, which is indispensable in current discussions on environmental issues. Otherwise, the new practices of the sustainable city could well be only the new clothes of the male domination" (author's translation). The geographer Edith Maruéjouls (2006) has demonstrated this very concretely by highlighting the gentrification of public spaces, which the sustainable city multiplies. Municipal spaces devoted to leisure activities, such as skate parks and multi-sports fields like city stadiums, are almost exclusively frequented by men. These "gendered" spaces institutionalize the male presence by massively building facilities that are a masculine symbol with higher attendances by boys and men. In this way, an unequal value is established (male facilities are more present), an unequal redistribution (public money is directed mainly towards masculine practices) and unequal access (girls and women have fewer places to practice their femininity). The challenge is therefore to know how to appropriate public space on an equal footing by neutralizing the facility and by qualifying the external space from the point of view of stereotyped gendered practices.

At the end of this journey, the city appears as a symbolic object of modernity, to the point of concentrating its characteristics and its tensions. First of all, the city is a project for the rationalization of space, of halting nature, which translates into the hold that human societies wish to have on their own destinies. We cannot help but refer to an episode from Genesis. Adam and Eve's eldest son, Cain, a sedentary farmer, commits the first murder of humanity by killing Abel, his brother, a nomad and the favorite of Yahweh (i.e. God). Condemned to wandering, Yahweh nevertheless guarantees his protection through the Ark of the Covenant, which always moves with the people. However, Cain cannot bear the uncertainty inherent in wandering and so founds the city of Enoch, which contains the idea of (re) "beginning". He surrounds it with walls, out of distrust of God's protection, so that he might defend himself. He invents technology in order to control the world, and thus widens the chasm between God and himself. The city fulfills his desire for security. From the start, the city appears as the project for a human Eden that substitutes that of God's, the environment created by humans to escape God's project. Cain, by building the first city, marks the beginning of humanity. At the same time, the questioning of the progressive paradigm and its techno-scientific references that affects industrial societies profoundly renews the way in which the city is thought of and built. To be sustainable, the city no longer relies on a theoretical catechism but on empirical, multi-actor and multi-level methods. In other words, the enterprise of the city changes its epistemological reference points, where a collective for empirical co-production must compensate for the theoretical cooling. The city thus needs social mobilization to regenerate itself, to find answers to the inflation of inequalities that it harbors and to confront the ecological crisis. However, the city also embodies the tensions of modernity as a theater for the development of individualism. The latter certainly has the virtue of emancipating individuals, of freeing them from the collective yoke. But, at a time when urban challenges seem to increasingly call for solidarity and cooperation, the city continues to separate, segregate and isolate, coming up against the difficulty of establishing a society of individuals. This does not mean that they are totally disengaged from the common good, but they do not find themselves in the traditional forms of civic engagement. If the reference to the village, the idealized community figure of the past, remains powerful, the future should result more from the invention and dissemination of practices that manage to consider the paradoxical desire of the city dweller, to be both free and engaged.

4.4. References

Aalborg Charter (1994). Charter of European cities & towns towards sustainability [Online]. Available at: https://sustainablecities.eu/fileadmin/repository/Aalborg_Charter/Aalborg_Charter_English.pdf [Accessed 25 February 2022].

Aquachar-Charpentier, M. (1997). Le Péri-urbain. Note de synthèse et bibliographie. Note, Centre de documentation de l'urbanisme, Ministère de l'Équipement, du Logement, des Transports et du Tourisme.

Ascher, F. (2005). *Métapolis ou l'avenir des villes*. Odile Jacob, Paris.

Augustin, J.-P. (2010). L'urbanité flexible des espaces publics. In *50 questions à la ville*, Augustin, J.-P. and Favory, M. (eds). Maison des Sciences de l'Homme d'Aquitaine, Pessac.

Béal, V., Charvolin, F., Morel Journel, C. (2011). La ville durable au risque des écoquartiers. Réflexions autour du projet New Islington à Manchester. *Espaces et sociétés*, 2011/4 (147), 77–97.

Beck, U. (1998). *World Risk Society*. Polity Press, Cambridge.

Berger, M., Rougé, L., Thomann, S., Thouzellier, C. (2010). Vieillir en pavillon : mobilités et ancrages des personnes âgées dans les espaces périurbains d'aires métropolitaines (Toulouse, Paris, Marseille). *Espace populations sociétés* [Online]. Available at: http://journals.openedition.org/eps/3912 [Accessed 08 January 2021].

Blondiaux, L. (2008). *Le nouvel esprit de la démocratie. Actualité de la démocratie participative*. Le Seuil, Paris.

Boltanski, L. and Chiapello, E. (1999). *Le nouvel esprit du capitalisme*. Gallimard, Paris.

Bordreuil, J.-S. (2000). La ville desserrée. In *La ville et l'urbain*, Paquot, T., Lussault, M., Body-Gendrot, S. (eds). *L'état des savoirs*. La Découverte, Paris.

Bourdin, A. (2001). Comment on fait la ville aujourd'hui en France. *Espaces et sociétés*, 105–106, 148–166.

Bourdin, A. (2005). *La métropole des individus*. L'Aube, Paris.

Boutinet, J.-P. (1990). *Anthropologie du projet*. PUF, Paris.

Callon, M., Lascoumes, P., Barthe, Y. (2001). *Agir dans un monde incertain. Essai sur la démocratie technique*. Le Seuil, Paris.

Chalas, Y. (1998). L'urbanisme comme pratique. Pensée faible et débat public. *Les Annales de la recherche urbaine*, 80–81, 205–214.

Chalas, Y. (2003). *L'invention de la ville*. Economica, Paris.

Chalas, Y. (2004). L'urbanisme dans la société d'incertitude. In *L'imaginaire aménageur en mutation*, Chalas, Y. (ed.). L'Harmattan, Paris.

Chalas, Y. and Soubeyran, O. (2009). Incertitude, environnement et aménagement, Quelle rupture ? In *Comment les acteurs s'arrangent avec l'incertitude*, Chalas, Y., Gilbert, C., Vinck, D. (eds). Éditions des archives contemporaines, Paris.

Charmes, E. (2005). *La Vie périurbaine face à la menace des gated communities*. L'Harmattan, Paris.

Charmes, E. (2011). *La ville émiettée. Essai sur la clubbisation de la vie urbaine*. PUF, Paris.

Charmes, E. and Saoumi, T. (2009). *Villes rêvées, villes durables ?* Gallimard, Paris.

Choay, F. (1965). *L'urbanisme, utopies et réalités*. Le Seuil, Paris.

Clark, W.A.V. (1991). Residential preferences and neighborhood racial segregation. A test of the Schelling segregation model. *Demography*, 28–1, 1–19.

Coulon, A. (2012). *L' École de Chicago*. PUF, Paris.

Denèfle, S., Bresson, S., Dussuet, A., Roux, N. (2015). *Habiter le Corbusier*. Presses Universitaires de Rennes, Rennes.

Desage, F. (2017). Les exclus de l'inclusion. Construire du logement social en temps d'austérité et de mixité (France–Québec). *Espaces et sociétés*, 2017/3(170), 15–32.

Didier, S. (1997). Disney urbaniste : la ville de Celebration en Floride. *Cybergeo: European Journal of Geography* [Online]. Available at: http://cybergeo.revues.org/1147 [Accessed 08 January 2021].

Donzelot, J. (2006). *Quand la ville se défait*. Le Seuil, Paris.

Ehrenberg, A. (1998). *La fatigue d'être soi*. Odile Jacob, Paris.

Ferguson, Y. (2011). Les conditions de gouvernabilité du développement urbain durable. In *Le développement durable changera-t-il la ville ? Le regard des sciences sociales*, Béal, V., Gauthier, M., Pinson, G. (eds). Presses Universitaires de Saint-Étienne, Saint-Étienne.

Ferguson, Y. (2014). Politiser l'action publique. Une approche par les instruments. Doctoral thesis, Toulouse II Le Mirail.

Germain, A. and Archambault, J. (1995). Cohabitation interethnique et vie de quartier. Final report, Ministère des affaires internationales, de l'immigration et des communautés culturelles and Ville de Montréal.

Golovtchenko, N. and Souchet, F. (2005). Des gated communities à la française ? Les résidences fermées toulousaines. In *La société des voisins. Partager un habitat collectif*, Haumont, B. and Morel, A. (eds). Maison des Sciences de l'Homme, Paris.

Grafmeyer, Y. (1994). *Sociologie urbaine*. Nathan, Paris.

Grafmeyer, Y. and Auhtier, J.-Y. (2008). *Sociologie urbaine*. Nathan Université, Paris.

Grafmeyer, Y. and Joseph, I. (2004). *L'École de Chicago*. Flammarion, Paris.

Jacobs, J. (1991). *The Death and the Life of Great American Cities*. Random House, New York.

Jaillet, M.-C. (1999). Peut-on parler de sécession urbaine à propos des villes européennes ? *Esprit*, 258(11), 62–85.

Jaillet, M.-C. (2001). Ces quartiers dont on parle. *Profession Banlieues, les 5 à 7* [Online]. Available at: https://www.professionbanlieue.org/Ces-quartiers-dont-on-parle. [Accessed 08 January 2021].

Jonas, S. and Weidmann, F. (2006). *Simmel et l'espace. De la ville d'art à la métropole*. L'Harmattan, Paris.

Joseph, I. (1992). L'espace public comme lieu de l'action. *Annales de la recherche urbaine*, 57–58, 210–217.

Le Corbusier (1925). *Urbanisme*. Flammarion, Paris.

Le Corbusier (1937). *Quand les cathédrales étaient blanches*. Bartillat, Paris.

Le Corbusier (1971). *La Charte d'Athènes*. Le Seuil, Paris.

Le Corbusier (1995). *Vers une architecture*. Flammarion, Paris.

Lees, L. (2003). Visions of urban renaissance: The urban task force report and the urban white paper. In *Urban Renaissance, New Labour, Community and Urban Policy*, Imrie, R. and Raco, M. (eds). The Policy Press, London.

Lewis, E. and Slitine, R. (2016). *Le coup d'état citoyen*. La Découverte, Paris.

Lorrain, D. (1995). La grande entreprise urbaine et l'action publique. *Sociologie du travail*, 37–2, 199–220 [Online]. Available at: https://www.persee.fr/doc/sotra_0038-0296_1995_num_37_2_2204.

Marchal, H. (2011). *Les grandes questions sur la ville et l'urbain*. PUF, Paris.

Maruéjouls, E. (2006). Loisirs des jeunes dans le secteur public : comment éviter l'exclusion des filles ? In *Mixité, parité, genre dans les métiers de l'animation*, Gillet, J.-C. and Raibaud, Y. (eds). L'Harmattan, Paris.

Maurin, E. (2011). La ségrégation urbaine, son intensité et ses causes. In *Repenser la solidarité*, Paugam, S. (ed.). PUF, Paris.

Merlin, P. (2009). *L'exode urbain*. La Documentation française, Paris.

Merton, R.K. (1938). Social structure and anomie. *American Sociological Review*, 3(5), 672–682.

Ministère de l'agriculture de la République française (2011). Les paysages agricoles dominent toujours le territoire français. Agreste Primeur [Online]. Available at: https://agreste.agriculture.gouv.fr/agreste-web/download/publication/publie/Pri260/primeur260.pdf [Accessed 05 April 2022].

Oberti, M. (2001). Le communautarisme de classe : distance spatiale et sociale comme alternative à la mixité sociale. *Mouvements*, 2001/3(15–16), 212–214.

Orfeuil, J.-P. and Solleyret, D. (2002). Quelles interactions entre les marchés de la mobilité à courte et longue distance ? *Recherche transport sécurité*, 76, 208–221.

Park, R. (1925). Propositions de recherche sur le comportement humain en milieu urbain. In *L'École de Chicago*, Grafmeyer, Y. and Joseph, I. (eds). Flammarion, Paris.

Park, R. (1928). Human migration and the marginal man. *American Journal of Sociology*, 33(6), 881–893.

Pech, T. (2011). Les inégalités menacent la cohésion sociale. *Alternatives Economiques,* Special edition no. 089 [Online]. Available at: https://www.alternatives-economiques.fr/inegalites-menacent-cohesion-sociale/00059723.

Pinçon, M. and Pinçon-Charlot, M. (1999). *Dans les beaux quartiers*. Le Seuil, Paris.

Pinçon, M. and Pinçon-Charlot, M. (2007). *Les ghettos du gotha*. Le Seuil, Paris.

Pinson, G. (2009). *Gouverner la ville par projet : urbanisme et gouvernance des villes européennes*. La Découverte, Paris.

Putnam, R. (2000). *Bowling Alone: The Collapse and Revival of American Community*. Simon and Schuster, New York.

Raibaud, Y. (2014). A Bordeaux (comme ailleurs), le vélo est au masculin [Online]. Available at: https://rue89bordeaux.com/2014/05/bordeaux-le-velo-est-au-masculin/ [Accessed 08 January 2021].

Raibaud, Y. (2015). La ville durable creuse les inégalités [Online]. Available at: https://lejournal.cnrs.fr/billets/la-ville-durable-creuse-les-inegalites [Accessed 08 January 2021].

Raymond, H., Haumont, N., Dezès, M.G., Haumont, A. (2001). *L'habitat pavillonnaire*. L'Harmattan, Paris.

Recoquillon, C. (2014). Néolibéralisation et (in)justice spatiale : le cas de la gentrification de Harlem. *Justice spatiale*, 6, June [Online]. Available at: https://www.jssj.org/article/neoliberalisation-et-injustice-spatiale-le-cas-de-la-gentrification-de-harlem/ [Accessed 08 January 2021].

Rosanvallon, P. (2011). *La société des égaux*. Le Seuil, Paris.

Rougé, L. (2010). L'installation périurbaine entre risque de captivité et opportunités d'autonomisation. *Articulo – Journal of Urban Research* 5 [Online]. Available at: https://journals.openedition.org/articulo/1440 [Accessed 08 January 2021].

Rudder, V. (2002). De l'urbain au social : le cycle des relations raciales. *Revue européenne des migrations internationales*, 18(3) [Online]. Available at: http://journals.openedition.org/remi/1607 [Accessed 08 January 2021].

Saoumi, T. (2009). *Ecoquartiers : secrets de fabrication : analyse critique d'exemples européens*. Les Carnets de l'information, Paris.

Saoumi, T. (ed.) (2011). *Ecoquartier un urbanisme durable*. La Documentation française, Paris.

Schnapper, D. (2005). Idéal et limites de la mixité sociale. *Informations sociales*, 5/2005 (125), 6–15 [Online]. Available at: www.cairn.info/revue-informations-sociales-2005-5-page-6.htm [Accessed 20 January 2020]

Simmel, G. (1972) [1903]. The metropolis and mental life. In *Georg Simmel: On Individuality and Social Forms*, Levine, D.N. (ed.). The University of Chicago Press, Chicago.

Simmel, G. (1976) [1908]. The stranger. *The Sociology of Georg Simmel*. Free Press, New York.

Stébé, J.-M. (2007). *La crise des banlieues : sociologie des quartiers sensibles*, 3rd edition. PUF, Paris.

Stébé, J.-M. (2009). *Mythologie des cités ghettos*. Éditions Le Cavalier Bleu, Paris.

Turner, M. and Wolman, H. (2006). Processus et politiques de ségrégation raciale aux États-Unis. *Hérodote*, 3(122), 44–65.

Weber, M. (1958) [1921]. *The City*. Free Press, Glencoe.

Weil, P.E. (2017). *Sans toit ni loi ? genèse et conditions de mise en œuvre du droit au logement opposable*. Presses universitaires de Rennes, Rennes.

Wilkinson, R. and Pickett, K. (2010). *The Spirit Level: Why Equality is Better for Everyone*. Penguin, London.

List of Authors

Luc ADOLPHE
Laboratoire de Recherche
en Architecture
École Nationale Supérieure
d'Architecture de Toulouse
France

Fabrice ESCAFFRE
LISST Laboratoire Interdisciplinaire
Solidarités, Sociétés, Territoires
UT2J, CNRS UMR5193
Toulouse
France

Yann FERGUSON
Icam
Toulouse
France

Mélanie GAMBINO
LISST Laboratoire Interdisciplinaire
Solidarités, Sociétés, Territoires
UT2J, CNRS UMR5193
Toulouse
France

Sinda HAOUÈS-JOUVE
LISST Laboratoire Interdisciplinaire
Solidarités, Sociétés, Territoires
UT2J, CNRS UMR5193
Toulouse
France

Patrizia LAUDATI
SicLab Méditerranée
UCA Université Côte d'Azur
Nice
France

Marc MÉQUIGNON
LERASS
UPS
Toulouse
France

Jean-Pierre MIGNOT
LERASS
UPS
Toulouse
France

Index

Printed and bound by CPI Group (UK) Ltd, Croydon, CR0 4YY
13/09/2026
14961330-0001